PHEASANTS

including their Care in the Aviary

PHEASANTS
including their Care in the Aviary

BY

H. A. GERRITS

Zoological Gardens, Wassenaar

Illustrated by H. J. SLIJPER

LONDON

BLANDFORD PRESS

Blandford Press Ltd.,
16 West Central Street,
London W.C.1.

Prepared from the original
Dutch publication
Handboek voor de Fazantenweker

*Printed in Great Britain by Richard Clay and Company, Ltd.,
Bungay, Suffolk*

CONTENTS

The Publishers gratefully acknowledge the help and advice readily given in the preparation of this English language edition by Mr. Terry Jones of Leckford Abbas and Mr. Philip Wayre of the Ornamental Pheasant Trust. Both Mr. Jones and Mr. Wayre, as well as the Zoological Society of London, have kindly loaned photographs which are separately acknowledged.

Foreword

MR. GERRITS' book will be widely read, as it is written by someone who knows about pheasants and has clearly described their management in captivity.

As the world's human population grows, there is less and less room for the wild animal and bird populations, especially if they happen to be species which cannot adapt themselves to cultivated countrysides, and, as in the case of the pheasants, if they also have the ill luck of being extremely edible. But the pheasants, with very few exceptions, take well to captivity, and so with care and some planning, stocks can be kept going in captivity, or better still where suitable environments can be found in a feral state.

I would urge anyone who is genuinely interested in preserving the rarer pheasants to specialize in a few species rather than a comprehensive collection, and to specialize in those species which are suitable to his climate and to hold about five pairs of each species, to mark the young of each pair and try to send out pairs which are not too closely related.

If Baron de Rothschild had not built a numerically strong stock from the only pair of Swinhoes imported in 1866, the species would not very likely survive in captivity today. The same is also true of the original Brown Crossoptilon importation. It seems to take a good many years to get a species safely established. Over 400 Temmincks Tragopans have been sent out from the Leckford Aviaries, yet possibly the loveliest and certainly the easiest in captivity of the Tragopans is by no means a common or secure species in captivity.

I hope this book will encourage more people to keep and breed this very attractive family of birds.

TERRY JONES

Leckford Abbas
June, 1961

Introduction

THIS book has been written in the hope of furthering a more general interest in those most handsome of birds, the Pheasants. It has not been possible to describe all the known members of the sub-family Phasianidae, and consequently those which are extremely rare have been omitted.

In the main those birds are included which are of importance to the Pheasant fanciers or are to be seen in Zoological Gardens and Aviaries; even so, it has been necessary to include quite a number of rare specimens. This may be to the good, for a better picture of a genus is often obtained when some lesser-known races belonging to it are mentioned as well.

There are extensive descriptions for those who want general information about the birds, and also for those who have the intention of including Pheasants in their aviaries. Both classes of reader will want to know something about the great variety in the Pheasant world. Those readers who desire to learn more about the Phasianidae will find a bibliography at the end of this book.

I often met with great difficulties in giving the colour descriptions. It proves next to impossible to give a clear picture of a bird of such an elaborate pattern as a pheasant in a few words. In some instances I have therefore referred to those given by M. Delacour in his magnificent and comprehensive book *The Pheasants of the World*, realizing that any effort to equal, let alone to surpass, this great expert on pheasants would be in vain.

Finally, I want to express my thanks to Mr. P. W. Louwman of the Zoological Gardens, Wassenaar, for all the knowledge and information he has made available to me.

H. A. GERRITS

General Information

PHEASANTS are found in greatly divergent types of country. For instance, some varieties live in humid, tropical lowland forests, others are found in the mountains at great heights; pheasants occur amid dry, hot plains and hills, and representatives of the sub-family are also found on the barren glaciers of the high mountains. Blood Pheasants, for example, live at heights of up to 15,000 feet, where they breed as well. Anything but a warm or agreeable climate can be expected at such altitudes.

Generally speaking, the Himalayas and western China may be considered the chief area of distribution of the birds. However, various members of the genus Phasianidae may be found near the Caspian Sea and all through Central Asia down to the main concentration points.

The building of the nest is done exclusively by the hen pheasant. The hatching of the eggs and the rearing of the young is also done without any assistance of the cock. Nevertheless, this does not imply that the pheasants live separately all the time. In some species the two sexes each go their own separate ways, and cock and hen only meet when they are urged by the approaching breeding season. Immediately after mating the birds part company again; notable examples are the Peacock Pheasant and the Argus. In other breeds a temporary and casual attachment occurs between the sexes, with a cock and one or several hens forming a group that stays together for a relatively short or long period. In no instance does pairing for life occur. As a result, it is hard to define whether the pheasant is a polygamous or a monogamous bird. Instances of both are found, and in addition the local density of distribution is an important influence.

When I wrote that the building of the nest is done by the hen this does not imply that she works very hard at it. As a rule the nest consists of a very simple hollow in the earth surrounded by some grasses, straws and twigs. Abandoned nests in trees or shrubs may be used, though this nesting habit is restricted to a few species and is mainly found with the Tragopans.

The number of eggs in a clutch varies—Peacock Pheasants and Argus Pheasants have only two eggs at a time (a peculiarity they share with the pigeons); all other species of pheasants lay four to twenty eggs in one period. When considering that the care of the eggs and the young is provided by the hen pheasant only, it should be pointed out that a similar situation occurs with all birds where the male has a highly ornamental plumage. I regard this as a safeguarding of the nest and the young, since such a brilliantly coloured bird would attract the attention of the natural enemies far too much to the nest and the chicks.

When the breeding season is over and the young have become self-supporting the birds unite into groups. Sometimes these are mixed flights; the old cocks form a separate group and the hens, with the young birds, assemble in another group. The number of birds in such a group and the period of time during which they remain together vary greatly. Eared Pheasants remain together almost all the year round, and are probably monogamous.

There are, however, pheasants that are more often found in pairs; in my experience this includes the Ruffed Pheasants, the Long-tailed Pheasants, Cheers and Tragopans. The breeds that unite in mixed groups for a relatively short or long period include the Peafowl, the Blood Pheasants, the Monals, True Pheasants and Gallopheasants. Argus Pheasants and Peacock Pheasants lose all interest in their mates once the young are raised.

The cock pheasant uses its ornamental plumage to attract the attention of the female. The bird often adds to its native splendour by very intricate display; these patterns will be further discussed in the individual descriptions of the various breeds of pheasants. Pheasants have short, rounded wings which enable them to fly fast but not for long distances. Of course, the various breeds have different flying capacities, and when in danger they show a remarkable difference in their manner of flight or in running away. Most pheasants sleep in trees.

Pheasants are Excellent Aviary Birds

Because the pheasant spends the greater part of its life on the ground, it is able to run fast and well. In the quest for food the earth is often scratched with the legs and/or the bill. Though pheasants are omnivorous, seeds and berries form the main part of their diet.

The pheasant is one of the birds that readily breeds outside its natural environment. With few exceptions pheasants are all good aviary birds and can be kept in full freedom even in a wood or park —provided the circumstances are favourable. It must, however, be pointed out that the different varieties have their own particular needs to be met in the way of climate, nature of the soil, etc. Every pheasant fancier should acquaint himself thoroughly with these demands so that he is able to meet the requirements of the birds as much as possible.

Feeding

ALTHOUGH the basic diet for the full-grown pheasant consists of cereals, this diet should be varied. It is advisable to provide a mixture of cereals, in which wheat is the main ingredient but which should also contain whole or broken maize, especially in autumn and winter.

During the breeding season the cereal mixture must be supplemented with various other seeds, such as buckwheat, millet, barley and corn; adding a little hemp seed may prove useful. If the necessary variety is omitted the breeder runs the risk that the birds, fed exclusively on cereals, grow too fat—which unfailingly results in clear eggs.

Variety in the feeding is a *necessity*, and pheasants fed on cereals at night (never give more than the birds can eat in one hour) must be given a mixed menu in the morning. This should consist of bran and oatmeal in varying proportions, to which have been added 5–10 per cent fish or meat meal, boiled potatoes, shredded carrots, vegetables, clover and lettuce. In Italy good results have been obtained with a mixture of silkworm cocoons, dried figs and maize meal.

Scientifically Prepared Mixtures

Nowadays it is possible to order feeding mixtures in balanced proportions from the large bird-food manufacturers. In this food all proprietary ingredients have been included in a scientifically prepared way, greatly simplifying the feeding problem. If possible, these mixtures should be given in preference to other foods.

In addition, a dry mash, with a protein content of 25 per cent, should be provided in the aviary, particularly during the laying period. Green food is essential. Every day a fresh quantity of lettuce, endive or other greenstuff should be put into the aviary, unless sufficient grass or other edible plants are available. Raw or cooked carrots and beets, berries and fruit are excellent as well, and for some species (chiefly Tragopan Pheasants, which are allowed few cereals) they are absolutely indispensable.

If there are newly imported birds in the aviary the food should in no instance consist of cereals only. Raw minced meat and hard-boiled eggs have already saved the life of many a bird not yet accustomed to a diet of seeds and are also indispensable in introducing the birds to a change of menu. In fact, these ingredients are a necessary part of the daily diet of Argus and Peacock Pheasants, even if they seem to be doing quite well on cereals alone.

In the same way some worms and insects should always be provided, but since these are sometimes hard to come by, they need not be given every day or in large quantities.

Those who wish their birds to have a treat may provide some nuts, bread, fruit and other delicacies, but these should be given in small quantities. If supplied within limits these feeds are very useful, but too much of them is quite harmful.

American Feeding Methods

Mr. George A. Allen Jr., Editor, *Game Bird Breeders' Gazette*, and Director of the Jean Delacour American Game Bird Park and Propagation Centre at Salt Lake City, Utah, has made the following contribution on American methods of feeding:

In America we follow a slightly different pattern. Some of our leading feed manufacturers have spent millions of dollars on research, and a good example is the Ralston Purina Company of St. Louis, Missouri. They have developed a well-balanced game-bird feed which contains everything necessary to grow strong, healthy game birds and other feed manufacturers also have developed game-bird feeds. Purina is a standard ration produced in a crumb-like consistency which is easily eaten by young chicks as well as older birds.

Pheasant production in America can be separated into two categories: the Commercial and the Ornamental.

The Commercial breeders are those who raise the common Ringneck types for meat purposes or for shooting (Ringnecks, Mongolians, Blacknecks, etc.). The Game Bird Industry is highly developed in the United States, and thousands of commercial breeders each raise as many as 10,000 to 100,000 birds a year. Incubators are used by commercial breeders in America with capacities of 40,000 or more eggs at one time, and often yield a hatchability of 85 per cent or more per hatch, depending on the quality of the machine and skill of the operator. Through selective breeding, Ringnecks, Blacknecks, or Mongolians, or crosses of these varieties, have been developed to weigh up to 7½ lb. While these meat birds are so big and fat they can hardly fly, the shooting-preserve pheasant (4–4½ lbs) is fast flying and as good a sporting bird in every way as those found in the wild state.

The other category of pheasant production in America is the Ornamental, made up of thousands of fanciers who rear the rarer types from the Goldens and Amhersts to the very rare and expensive Argus, Tragopans, Peacock Pheasants, etc.

But the feeding programme for all types of pheasants is very similar. They are all fed a prepared game-bird feed. Purina manufactures a starting feed or ration (they call it Startena), a finishing ration (Finisher), and a laying ration for adults and breeding stock (Layena).

Newly hatched chicks are fed Startena for about the first 45 days because it is extra rich in protein as well as other food elements so necessary. Also fresh, clean water and grit daily are necessary. Hard-boiled egg yolk should be finely mashed or grated and sprinkled on top or mixed in the game-bird food for the first few days of life. There is a very fine supplement called "Liv" which contains all the nutrition of the egg yolk and much more and is also a golden colour; it is in a dry, powdery form which does not spoil or sour easily, can be stored, and is therefore easier to feed than egg yolk. The colour of egg yolk or of a supplement attracts the attention of newly hatched chicks and helps to start them eating.

Eggs or supplements are usually discontinued after a week or so for common types of pheasants, and the prepared food is all that is necessary from then on. After about forty-five days the ration is changed to Finisher (a feed with less protein and less expensive), which is fed until the birds are fully mature. Cracked corn and wheat can be added to the diet after the first month or so. Some breeders feed finely chopped-up greens from the day the chicks are hatched, but many feed manufacturers claim the nutritional value of greens is included in their ration and is unnecessary.

However, with rearing the semi-rare and rare varieties, such as Elliot's Mikados, Coppers, Edwards, Firebacks, Tragopans, Peacock Pheasants, Brown and White-Eared (not Blue Eared), Argus, Koklus etc., I advise greens be added to the diet (finely chopped lettuce, endive, parsley, watercress and clover), and fed throughout

Plate I

Grey Peacock Pheasant
Polyplectron bicalcaratum

Southern Caucasian Pheasant
Phasianus colchicus colchicus

Southern Green Pheasant
Phasianus versicolor versicolor

Chinese Ring-necked Pheasant
Phasianus colchicus torquatus

the growing period. The common pheasants: Goldens, Amhersts, Reeves, Ringnecks, Swinhoes, Silvers Kalij etc., need not have this special attention if a good game-bird feed is used. Impeyans can also be included in the feeding programme of the common types, since I have found they do well on a simple diet, and I have raised as many as sixty a year on nothing more than Purina feed. Also the Blue Eared thrive on this diet.

Getting back to the rarer species, in addition to greens live food should be provided. Mealworms are the popular form of live food, and two, three or more worms per pheasant per day is essential for the successful rearing of many of these rare species. Also the continuance of a supplement such as Liv throughout the growing period is very good (from hatching to maturity). A supplement can even be fed the common types throughout the growing period if a richer diet resulting in faster growth is desired.

Sanitation is of the utmost importance, and the best feeding programme cannot grow healthy, strong birds unless good sanitary conditions are maintained. Clean water and grit, and clean feed containers are a "must" at all times. When brooded by bantams, mother and chicks should be moved to clean surroundings frequently.

As to feeding adult pheasants, a feed such as Layena can be used as a basic diet and kept before them most of the time, especially a few months before and during the breeding and laying season. Also a handful of grain, such as corn, wheat milo or others, can be thrown in the pen every other day. Adult birds should be kept in clean, dry pens, and sand is ideal for pen floors. Greens should be generously fed to adult ornamental pheasants. Live food should be given to the rare species during the breeding season in addition to their prepared feed and supplement. Some species show a liking for grapes, apples, pears, oranges and other fruit, and these should be included in the diet if possible, but are not absolutely necessary. Some grain can be fed during the laying season, but not too much, for pheasants should be encouraged to eat their prepared ration during this period. A rich, nutritional, non-fattening diet is best for adults, especially during the breeding season. A supplement such as Liv mixed in the ration increases the production ability of common and rare pheasants, but is especially valuable for the rarer types.

The more nutrition pheasants get, the better the propagation results. The feeding programme for the breeding stock determines the fertility and quality of the egg and whether the embryo will be strong enough to develop properly and the chick break out of the egg at the right time and begin eating normally. The feeding programme for the chick determines whether it will continue to live after hatching and develop into a healthy, normal adult. Purina Research scientists had the following to say in the *Game Bird Breeders' Gazette* on this subject:

> "Successful game bird breeders know it pays to build strength and vigour into game bird chicks long before the eggs from which they hatch are laid. Feeding both cocks and hens (breeding stock) a good laying feed well in

advance of hatching season provides high-quality proteins, extra vitamins, minerals, and other nutrients game birds need, not only for high production of fertile eggs, but for profitable hatches of strong, vigorous chicks off to a fast, safe start. Properly nourished long before they see the light of day, embryo-fed chicks outgrow many troubles before they strike."

In other words, the food fed to the breeders determines the quality of the embryo and newly hatched chick.

The following table shows the ingredients which go into the three types of Purina's scientifically prepared foods, together with the analysis. Prepared game-bird feeds of other manufacturers' in America, and probably in Europe, have a similar analysis.

All of the thousands of pheasant fanciers and commercial breeders throughout the United States now use a prepared feed. It is interesting to note the higher percentage of protein used for pheasant feeds than is used for domestic fowl.

	Ingredients	Analysis
Game Bird Startena	Fish meal, meat and bone scrap, dried whole whey, soybean meal, dehydrated alfalfa meal, ground barley and/or ground oats, ground yellow corn and/or ground grain sorghums, wheat middlings, condensed fish solubles, animal fat preserved with BHA (butylated hydroxyanisole), BHT (butylated hydroxytoluene—a preservative), brewers' dried yeast, vitamin B_{12} supplement, procaine penicillin, riboflavin supplement, calcium pantothenate, niacin, choline chloride, vitamin A oil, D activated animal sterol, vitamin E supplement, menadione sodium bisulfite (source of vitamin K activity), 1·5% low fluorine rock phosphate, 1·5% calcium carbonate, 0·5% iodized salt and traces of manganese sulphate, manganese oxide.	Protein 30·0 Fat 3·0 Fibre 6·5 Sulfaquinoxaline 0·0175
Game Bird Finisher	Ground yellow corn and/or ground grain sorghums, meat and bone scrap, fish meal, soybean meal, dehydrated alfalfa meal, ground barley and/or ground oats, wheat middlings, animal fat preserved with BHA (butylated hydroxyanisole), condensed fish solubles, vitamin B_{12} supplement, procaine penicillin, riboflavin supplement, D activated animal sterol, vitamin A oil, niacin, 1·5% low fluorine rock phosphate, 2% calcium carbonate, 0·5% iodized salt and traces of manganese sulphate, manganese oxide (SQ—5429).	Protein 20·0 Fat 2·5 Fibre 6·0
Game Bird Breeder Layena	Meat and bone scrap, ground barley and/or ground oats, fish meal, soybean meal, ground yellow corn and/or ground grain sorghums, dehydrated alfalfa meal, dried whole whey, wheat middlings, condensed fish solubles, BHT (butylated hydroxytoluene—a preservative), vitamin B_{12} supplement, riboflavin supplement, vitamin A oil, D activated animal sterol, vitamin E supplement, menadione sodium bisulphite (source of vitamin K activity), 1% low fluorine rock phosphate, 3% calcium carbonate, 0·5% iodized salt, 0·02% and traces of manganese sulphate, manganese oxide.	Protein 20·0 Fat 3·0 Fibre 7·0

Housing

APPROPRIATE housing for a pheasant couple or a cock and a few hens consists of a night house and a run. Suitable dimensions for a night house are: depth 5–10 feet, and width 7–12 feet. The run should be 9–12 feet wide and 12–25 feet long.

FIG. 1. The kind of surroundings which are necessary in the planning of a Pheasant Aviary.

Both the house and the run must be high enough to allow a grown man to stand upright easily. A height of 7 feet is recommended.

For the larger varieties of birds smaller dimensions are too restricted. It must be remembered that pheasants thrive in proportion to the space available; hence the more room the better.

Building a Night House

The night house should be of wood, with a concrete, stone or wood floor. Care must be taken to ensure enough daylight, and sturdy perches should be provided. Baskets, small crates, sods and such like should be installed for nesting accommodation and cover.

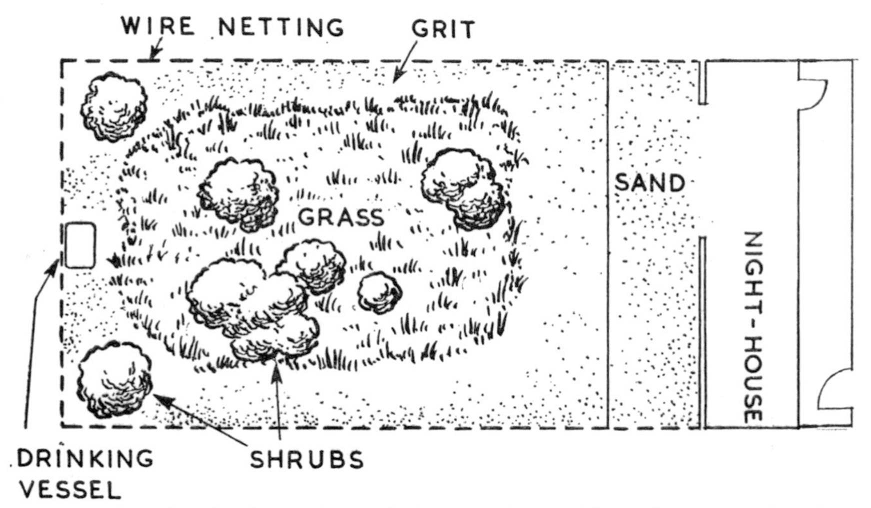

FIG. 2. Plans showing layout for night-house and run with shrubs, grass and sand. The run can, of course, be made to any length.

Most species of pheasants do not require artificial heat. After one year even tropical breeds should have become sufficiently acclimatized to stand cold winters without harm, provided that they are shut in at night and during bad weather.

Pheasants are much more sensitive to wind and damp than to cold. It is essential always to check the whereabouts of the birds, as they prefer sleeping out of doors, and this can have harmful consequences. It is also wise to divide the night house into two parts so that it is possible when necessary to lock up both sexes separately without their having to be caught.

For all hardy pheasants (these are described in separate chapters) the night house, if preferred, can be built with an open or a wire-mesh front, provided with a door. For the less hardy varieties an enclosure made of glass or a plastic is sufficient. When glass is used as a fence it is advisable to protect it with wire mesh on both sides or to whitewash the glass so as to avoid a frightened bird flying through the panes. Care must also be taken to ensure sufficient ventilation in a closed night house to avoid excessive heat, especially in summer.

Wintering

In a temperate climate non-hardy pheasants can be allowed to winter safely without artificial heating. The ground, however, should be covered with a thick layer of straw. By removing the perches the birds can be forced to sleep in the straw, so that their toes (the weakest spots in a pheasant) cannot freeze. If the night house is fairly large and the winter very severe it is advisable to surround the entire house with pallets or to install a heating lamp to keep the temperature just above freezing point, especially at night. The drinking-water may be placed above a floating wick, as shown in the drawing.

The Run

The run is preferably made of fine-meshed wire netting, stretched over a wooden or iron framework. If the aviary is placed on a low brick wall this should be sunk to a sufficient depth to prevent rats, mice and other vermin from entering the cage.

If, on the contrary, no foundation is laid, the fine-wire mesh should be dug deeply into the earth, which also provides good protection for a long time. To safeguard the birds from cats, it is advisable to install either a double wire-mesh roof (spaced at about 4–5 inches) or an electric wire fence. Part of the run should be closely planted with shrubs, low trees or bushes; this makes the pheasants feel much more at ease and ensures better breeding results.

When a consecutive row of aviaries is built with communal walls it is better to plant these with a hedge to a height of 25 inches above the ground or to cover them with a wooden partition. This prevents

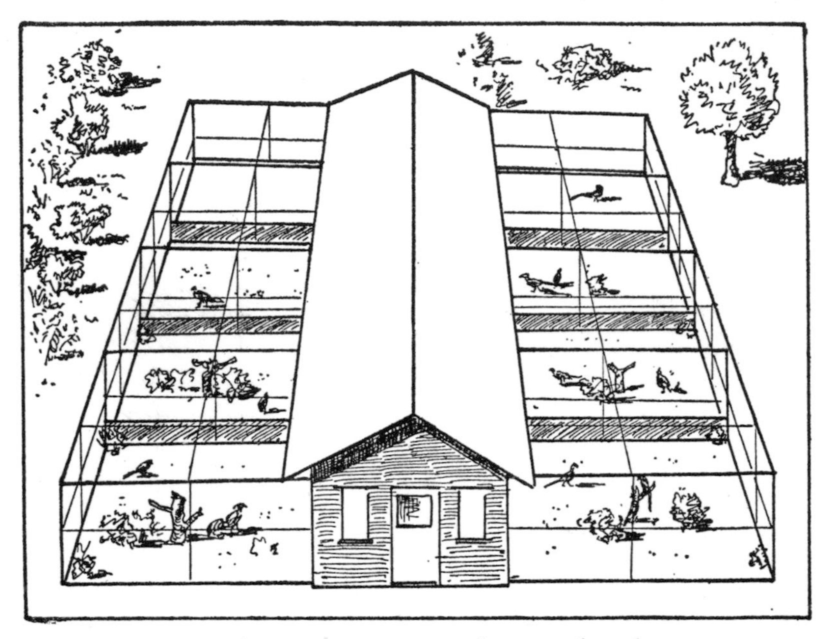

FIG. 3. A series of runs similar to Fig. 2 incorporated into one large house; a gangway runs down the middle of the house with the aviaries and their roosting places on each side.

fighting across the wire mesh and also gives the birds more restful surroundings. But if care is taken to see that no pheasants of the same variety, or closely related varieties, are housed side by side, this division will usually be unnecessary.

A hedge of low privet or buxus makes an attractive bordering for the aviary, and when the corners are planted with larger and thicker bushes the birds, who like to walk closely along the edges, will not damage their tails by being forced to "square" the corners. They will circle round instead.

The Floor of the Aviary

The ground in the run must be of a loose consistency, slightly sloping and well drained. Turf is strongly recommended, with a bird-walk provided along the sides.

Grit or ground shells are indispensable in an aviary and, in addition, a spot of dry sand or fine dry earth for an "earth bath" should be provided.

As a matter of course every aviary should contain a large flat water-dish, which must be carefully kept clean. Running water is not recommended; it prevents disinfectants and medicines from taking effect.

If sufficient space is available a wire-enclosed walk can be made in front of the aviaries, with the separate doors opening on to it. Not only does this make escape impossible, but it also ensures that the birds can be moved easily from one house to the other without having to be caught.

While the size of the aviary described at the beginning of this chapter may be regarded as ideal, it will often not be possible to provide so much space for the pheasants or to construct such elaborate night houses. Fortunately, many species can be kept in much smaller spaces, and most of the hardy birds can safely be left outside, even in severe winters, with no more shelter than a screened corner.

Some Species Make Special Demands

The larger runs, sheltered against wind and fierce sunlight, will always be preferable to small and exposed ones. For Tragopan and Argus Pheasants the former are essential. In fact, all the bigger pheasants should be housed as spaciously as possible if their plumage is to be kept in a good condition. Every need of the birds should be amply met. If an aviary with a nicely planted lawn is desired, it is advisable to cover the grass with fine-meshed wire. The grass will grow through the meshes so that the wire becomes completely invisible. The pheasant will not be bothered by the wire mesh when walking in the run, but, at the same time, it will not be able to dig furiously with bill and feet, and the turf will benefit.

In the accommodation of the pheasants some very important points should be kept in mind. We have already seen that the pheasant has a very extensive area of distribution, and in a wild state may be found both in tropical lowlands and in the cold of the high mountains. Of course it is impossible to meet the greatly divergent needs for every bird separately. Consequently we will have to find a site for our aviaries that approaches as nearly as possible to the requirements of the birds and also will not harm their health. Practice has shown the best siting to be facing south or east. Strange though it may seem, pheasants from the Himalayas support a southern aspect in summer much better than a northern aspect in winter. Needless to say that the same rule applies to birds from hot tropical areas, and consequently a northern or western aspect should be avoided at all times. It should be remembered, however, to provide sufficient shade in the aviaries during the summer months; this holds good for all pheasants wherever they come from. Pheasants avoid as much as possible the direct rays of the sun. The best solution would be—if sufficient space is available—to provide a central night house with runs with a northern aspect to be used in summer and for the winter months runs with a southern exposure.

One of the unfortunate characteristics of pheasants is their unconquerable shyness. It has happened many times that pheasant fanciers lose their birds through lack of caution when approaching the aviaries. This causes the pheasants to fly up recklessly, damaging their skulls against the roof, which usually proves fatal. If a pheasant has torn its scalp it is best to stitch the wound immediately, the sides being brought together as closely as possible. Unless this is done the bird will have a bald skull for the rest of its life, provided it survives the wound and the risk of infection.

But prevention is better than cure, and often breeders will clip a number of flight feathers from one wing, or even clip both wings. Clipping is particularly important in recently imported birds. On the other hand, care should be taken that the hen will not be the victim of the aggressiveness of the cock. When the hen is deprived of her flying power she is also robbed of her only means of escape in a fight. Consequently the mutual tolerance of a mated pair should be taken into account. If one does not care to risk clipping the wings of the hen another means of preventing the bird from killing itself in sudden flight is the installation of a strong, fine-meshed net about 10 inches below the wire netting of the roof of the aviary. Of course the meshes have to be very fine so the bird cannot possibly hang itself in them.

In a Pheasant House

As this book will show, pheasants are combative birds. Usually, not more than one pair, or a cock with two or three hens, can be housed in the same aviary. But occasionally a number of cocks—*without hens*—may be successfully accommodated in one space, though fights will still occur from time to time. But, as a rule, birds of different species do not fight, and thus two or three varieties of different size and habits may easily be kept together in the same aviary if it is a large one.

Pigeons, parakeets and small birds make good company in a pheasant aviary, though, of course, a specialized food cannot then be provided, as most hoppers or troughs are accessible to all birds.

In limited spaces it will happen only too often that pheasant cocks molest or even kill their hens. Large rooms, hiding-places formed by shrubs, clipping of the wings, high and low perches help to prevent this, as do the putting on of spectacles or the shortening of the beak. In extreme cases, or when no peace-loving cock is available, the only solution is to house the cock and hen separately, though alongside each other. They must be brought together— and then under constant supervision—only when the hen shows obvious signs of being in breeding condition.

All races of the species Syrmaticus, especially the Copper Pheasant, show to a greater or lesser degree this bad habit, though such excesses may also be expected in all other pheasants. The reason is usually because both sexes do not enter a breeding condition at the same time; this, in turn, can be traced largely to the composition of the diet.

Getting Used to the Aviary

Special precautions should be taken when pheasants are first put into a new aviary. There should be nothing to frighten them, and the surroundings must be quiet. For wild or shy birds it is essential that the aviary should contain bushes or shrubs in which to hide. In the formation of pairs, or when a cock is brought together with a hen for the first time, care should be taken to see that the weakest—usually the hen—feels completely at home in the aviary. The best result is obtained when the hen has been living in the aviary for a few weeks before the cock is put into it. The latter takes some time to feel at home and, unaccustomed to his new surroundings, will leave the hen in peace for a while.

It should not be necessary to add that the quarters must be kept spotlessly clean. The night houses need to be disinfected thoroughly at least once a year; gravel walks and unplanted areas in the aviary must be disinfected regularly with a chloride, chalk or ordinary kitchen salt. This kills earthworms which, especially in the last few years, have become an ever-growing danger by their transmission of *Synchamus trachealus* (gape worm).

Brooding and Rearing of the Young

IN the month of February is the time to start making a definite choice of pairs to be mated, if necessary placing the birds in special breeding houses. As the first eggs may be expected in March, it would be fatal to mate the pairs later than this. The pheasant breeder who keeps his birds in the same space all through the year is advised to check his aviaries thoroughly once more to test their suitability for breeding.

I have in mind the possible planting of new "breeding shrubbery", improving the bushes that are to provide shelter, providing special nests if necessary, etc. Such plantings may be quite simple, consisting if need be of twigs of evergreen placed closely together. The only point is to provide an opportunity for the hen to lay her eggs in a sheltered spot. No need to exaggerate, or place the twigs so closely together that they shut out all the light. Hen pheasants never use breeding space that is too dark. Finally, a small hollow is dug in the spot designated for nesting. If possible several nests should be made so the hen can make a choice. For the breeding of Tragopans it is preferable to provide a nest above the ground, because in a wild state these pheasants usually nest in trees or shrubs. I would never advise the use of nesting baskets, as these make daily checking of the nest almost impossible.

In spite of all precautions, it often happens that the pheasants lay their eggs not only in the chosen nest but all over the aviary. These eggs should be collected and put into the nest until the clutch is complete.

Once the hen has begun to lay her eggs special attention should be paid to prevent fights with the cock. The latter will be considerably less aggressive during the breeding season, but if the cock interferes too much with the brooding it will have to be taken from the aviary. The same applies to several hens in one aviary. Even if they are all brooding, it may still be wise to separate them and to have the eggs hatched by Bantam hens or to use an artificial incubator. In this way there may be a risk of losing a few clutches of eggs, but, on the other hand, all eggs may be lost if they are left with the hen pheasants.

Still, hen pheasants are quite capable of hatching their own eggs. The rearing of the chicks can be safely entrusted to them as well. I would even go so far as to say that having the hen pheasants brood and rear the chicks themselves is *the* best way for breeders who have only limited time to spare for their birds. Of course the birds should not be left to themselves completely, and a daily check combined with good mashes is a must. Of course, this "easy" method has its drawbacks. In the first place, there is no second laying because of the long period of hatching and rearing, and secondly, quite a few chicks may be lost if the weather is unfavourable.

Brooding with a Bantam or a Small Hen

With a view to the above-mentioned drawbacks many breeders prefer to have their eggs hatched by a bantam or small hen. The size and weight of the foster-mother should correspond with those of the hen pheasant. Up till now the best results have been obtained with Japanese Silkies. These birds are easily induced to brooding, and in addition are very trusty broodies.

If this type of breeding is chosen, the eggs should be collected every day and marked on one side. While being hatched, the eggs should be carefully turned every day, and a few artificial or bad eggs should be put into the nest of the pheasant hen so the bird will not leave its nest. As soon as the clutch of eggs is complete the foster-mother is placed on the eggs.

After some time the pheasant hen will start laying a second clutch of eggs, and thus increase the chances of a sizeable number of offspring.

So as to safeguard the pheasant chicks as much as possible the breeder must make sure that the brooding hen is in good health and free from vermin.

It does not matter so much where the broody hen with the eggs is placed. If one has a garden it is preferable to have the brooding done out of doors. There are no objections against brooding indoors, provided the proper degree of moisture is ensured by means of spraying.

Excellent results are obtained from placing the hen with the eggs in a covered box or coop to which a small run, equally covered with a mesh wire-netting roof, has been attached. By sinking this box or coop a little way into the earth a sufficient degree of moisture is ensured in a very simple manner. The brooding hen has to be given an opportunity to feed and drink every day—it is very practical to place the food in the small run—and at the same time the breeder can check the eggs.

A day or two after the eggs have been hatched—this depends on the breed of the pheasant—the hen and the poults are transferred to a coop which can be closed and which is connected to a portable wire run. These runs have no floor—where there is danger from rats or other vermin the run should be fitted with a mesh-wire-netting floor—and must be moved to fresh ground twice a day. The turf must be disinfected carefully every year so that it does not become tainted with harmful disease germs.

Rearing of Young Pheasants

The chicks should be fed four to six times daily. There are many suitable types of chicken feed; much must depend on local conditions and on the particular variety of pheasant. It may be helpful to mention briefly the ingredients used by Delacour, well known as a successful breeder of pheasants. During the first ten to fifteen days the chicks were fed on fresh ants' eggs or cleaned flies' maggots. Mealworms were rarely used by Delacour, though other breeders have obtained excellent results, particularly with cut mealworms. The chicks were then fed on a mash consisting of eggs and milk well boiled together, to which were added, after cooking, biscuit-meal, millet seed and vegetable matter, according to season. As the days went by the chicken feed was changed and raw minced meat, dried ants' eggs and flies, and gradually a good mixture of biscuit and meat meal, with various cereals and seeds, were substituted.

It is extremely important to remove continually the food scattered and left behind in the run, so as to prevent the chicks eating spoilt food. After a few weeks the chicks may be allowed to run free on a lawn in the daytime; they then have to become accustomed to eating and sleeping in their night houses and, next, in a larger space. This method is no doubt the best, but unfortunately proves almost unworkable in practice. In any case, during the first six weeks, young pheasants should be kept in the small night coop and the small portable runs, after which they can be transferred to permanent aviaries, as spacious as possible. In these they become gradually accustomed to normal pheasant diet.

Good Mashes Available

In the course of the years scientifically composed mashes ready for use have been put on the market. These foods, used according to the instructions enclosed, have given very satisfactory results and have proved suitable for the rearing of the average pheasant. They avoid resort to expensive ingredients, such as fresh ants' eggs and dried insects, which are hard to get.

The more difficult species, however, require fresh insects during the first days to induce them to eat. Selected foods, such as the chicken food with boiled eggs described above, are also necessary. But now that a plentiful supply of reliable factory-made mashes is once more available, much work is saved to the pheasant breeder and the rearing of pheasant chicks is made simpler. Should time and space be limited, the poults of different varieties can be reared together, provided they are of the same size and do not belong to the combative breeds. It is always preferable not to mix different hatches. At the end of the brooding season the young pheasants are divided by sex, and new pairs, if any, are formed. Those pheasants that reach full maturity in their second year can be kept together, in large spaces, only until the next year.

Artificial Incubation

Eggs may, of course, be hatched by incubator, instead of by pheasant hens or broody hens. Modern machines are often so ingeniously constructed that they produce maximum results with the minimum of care. They also avoid the risk of transmitting contagious diseases.

It is not possible in this context to give directions for the use of incubators. Each machine has its own instructions; the many different makes render it impossible to give a general rule.

Incubators are supplied in every size and at all prices. When buying a machine it is always advisable to bear in mind that it pays in the long run to get the best. The simple types of incubator are restricted to the heating of the eggs; all such things as turning, ventilation, maintaining the required degree of moisture, etc., have to be done personally. The more complicated, and consequently the more expensive, machines regulate all these processes automatically, and require less experience from the user. But with precision and care most pheasant eggs may be successfully hatched in the incubator.

Some Hints about Artificial Incubation

If the reader possesses an incubator, or intends to make one to some definite design, the instructions given for the particular type of machine should be carefully adhered to when operating the incubator. It is advisable to try out and adjust the machine before putting in the eggs. While it is not possible to include in this book directions applicable to every incubator, some hints of a general nature can be given.

In most cases the eggs will be purchased from breeders or pheasant-keepers who advertise in the trade journals. It is, of course, difficult for suppliers to guarantee the fertility of their eggs. This must always be a question of trust, and it is therefore advisable to buy the eggs from somewhere near, and not to order them from a distance. A study of the prices for full-grown pheasants shows why the eggs of expensive varieties are often costly as well as difficult to obtain. The owners are usually breeders themselves, and it is more profitable to sell young pheasants at high prices than to sell eggs.

When selecting the eggs it is advisable to find out the date of laying, for eggs that are more than ten days old should never be bought. The shells should have a soft sheen; if they are dull and dirty and show spots they had better be left alone, because they will probably prove infertile or bad. If there is an opportunity to test them (it is easy to make a test lamp by placing a light-bulb in an empty tin and making a hole in the bottom, big enough to stand an egg on), it can be seen if the eggs are clear, without blood clots. Attention should also be paid to the air-chamber, which must not be at the wrong end or be excessively large. Eggs with a rough and bumpy surface should preferably be avoided.

In an incubator the correct degree of moisture must be maintained. A shallow dish of water should be replenished regularly. It is preferable to buy a hygrometer, and to ensure that the amount of moisture remains between 45° and 50°. It is fatal to incubate the eggs

Plate II

Black-breasted or Horsfield's
Lophura leucomelana lathami

Lineated Pheasant
Lophura leucomelana lineata

Black-backed Pheasant
Lophura leucomelana melanota

Nepal Pheasant
Lophura leucomelana leucomelana

in an atmosphere containing too much moisture. This will result in eggs becoming "dead in the shell". On the other hand, too dry an atmosphere will cause dessication of the germ.

It is important to know whether the hens have been kept in an outside run or in an inside aviary where they have been fed on exclusively dry mashes. The eggs in the latter case will be definitely inferior to those of birds that have had access to all useful vitamins and minerals, and especially to green food. All these useful ingredients will be found in the eggs. When the parent birds have had a varied diet the results will be evident in the young.

The brooding temperature is of major importance. As the eggs are placed side by side under lamps, and so are not in direct contact with the source of heat, a rather high brooding temperature, in the region of 104° F., should be maintained during the first four days. After that it can be reduced to 101° F., to be increased to 104° F. again for the last three days before hatching.

As a rule the degree of heat should be regulated by a thermostat. Instructions for use are supplied with the instrument. It may stimulate hatching if the amount of moisture is increased as soon as the eggs begin to chip. Sprinkling with lukewarm water, however, is better avoided. The eggs should be placed closely together so as to prevent loss of their own warmth. They should be turned in the morning and at night, when they will, at the same time, be ventilated. It is advisable to mark the eggs with a cross on one side, so that their correct position can be checked. The ventilation must be well regulated. Eggs should have a regular supply of fresh air, such as is provided when they are turned. They should never be allowed to cool so much that they become actually cold. This is especially important during the first week, when they should be aired for only a very short time. Later on there is less danger, as they will radiate warmth themselves owing to the development of the chick. As the eggs continue to develop, more air may be admitted. It is a good plan to change the position of the eggs on the outside with those in the middle some time during the incubation period. Any draught while the eggs are turned will prove fatal.

A manufacturer of various types of incubators has introduced a foam rubber incubator, which he calls the "electrical broody hen". Every pheasant fancier should be interested in such an apparatus, which is made in two sizes, one for holding thirty-five and the other for fifty-six pheasant eggs.

The eggs are placed in the foam rubber nest. The foam rubber retains the oxygen as well as the heat and moisture, so that it is rarely necessary to add moisture during the incubation period. A thermometer is placed on top of the eggs and everything is then covered with a plastic heating cushion in which has been installed the adjustable heating control, thus making the apparatus easy to operate. The electricity consumption is very low, viz. eight watts per hour. In an incubation period of twenty-one days only four kilowatts are used.

Good breeders will be interested in an automatized engine-driven incubator. In an incubator of this type between fifty-six and three-hundred pheasant eggs can be hatched, according to its size, as well as turkey eggs and also grouse eggs. The eggs are turned mechanically, so that little time need be spent on the brooding process, and the results will be nearly 100 per cent.

The great disadvantage of an incubator is, the lack of a mother for the chicks. Artificial brooders need a great deal of care and time.

HATCHING TIME-TABLE

The following hatching times for the various species of Pheasants, have been furnished by the American Game Bird Park and Propagation Centre, Salt Lake City, Utah, U.S.A.

	Days		*Days*		*Days*
Amherst	23–24	Formosan	24–25	Mikado	27–28
Bel's	26–27	Golden	23–24	Mongolian	24–25
Blackneck	24–25	Imperial	24–25	Nepal	24–25
Blue Eared	27–28	Impeyan	27–28	Reeves	24–25
Brown Eared	27–28	Kalij, Black-backed	24–25	Ringneck	24½
Cheer	26–27	Kalij, Lineated	26–27	Silver	26–27
Edwards	24–25	Kalij, White-crested	24–25	Swinhoe	24–25
Elliot's	24–25	Melanistic Mutant	24–25	Tragopans	27–28
Firebacks	24–25				

The Artificial Brooder

When all chicks have been hatched the young pheasants should be kept in the incubator until they are completely dry. It is even better to keep them for another twelve to twenty-four hours in the nursery section which is a part of most machines. They will not need food during this period, since just before they leave the egg they have absorbed the remains of the yolk through the navel. This nourishes and sustains them for some time. To provide food before these remnants of the yolk have been digested would be dangerous and might cause intestinal upsets.

To let the chicks drink they should be taken in the hand and their beaks dipped in a small dish of water for a moment. This enables them to absorb enough liquid. A saucer may also be used to provide water, but the chicks must not be allowed to get wet.

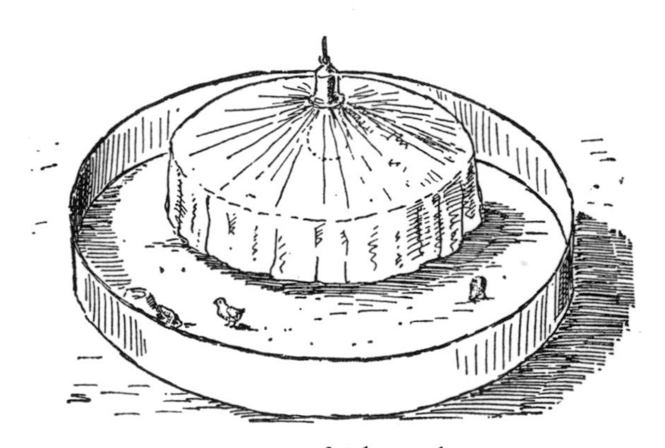

FIG. 4. Artificial Brooder.

The artificial brooder can be a chicken lamp or dark lamp, around which a cloth, reaching almost to the ground, has been hung. Dry, heated sand covers the floor. The hand is placed on the ground, and if after some time the heat is still easy to bear, then the lamp has been hung at the right height. In any case the chicks will move away from underneath the lamp if it becomes too hot for them. In that case the lamp should be suspended a little higher. The trays for food and drink should be fitted outside the radiation area so that the chicks are forced to run to and fro and to return under the lamp. On the ground under the lamp the temperature must not be allowed to exceed 105° F. The infra-red heating lamp which has nowadays taken the place of the carbon-filament lamp has much to recommend it; also it does not need a hood. The chicks feel very comfortable under it and can move freely in and out of the circle of warmth.

It depends on the site of the artificial brooder whether it is necessary to place a cardboard screen round the lamp to keep out draughts. If there is any draught at all such a screen is essential. Poults need a great deal of fresh air, so an upper window should be kept open day and night.

After a few days the chicks should be allowed to move about in a run connected to the coop with the artificial brooder. When the weather is wet or cold they should be kept inside; in dry weather they enjoy a stretch of turf and the sunlight. Artificial brooding involves quite a lot of work. The pheasant mother or dwarf hen considerably lightens the task of the breeder. Hence it is advisable for the inexperienced breeder to brood with hens or pheasant hens in the first years.

Rearing the Chicks

Naturally, many methods are used by different breeders in rearing their pheasant chicks, and it would be impossible in this book to describe them all. Readers would be overwhelmed by such a variety of good advice that they would not see the wood for the trees. But it may be helpful to quote C. F. Denley, one of the foremost breeders in the United States. In his excellent book *Ornamental Pheasants* he writes:

"There is a diversity of opinion among breeders as to the rations to be used in feeding the young. But here again there are certain fundamentals which must be observed, in order to get the best results. It is the aim in feeding poults to obtain the maximum of growth with the minimum of mortality. Just prior to hatching, a certain quantity of food substance is absorbed. Therefore, the poults do not require feeding until they are twenty-four to forty-eight hours old.

"For the first three weeks the poults should be fed four times daily; for the next three weeks, three times daily; after that, twice daily. They must be fed regularly—not gorged in the morning then starved until evening.

"Food for the young pheasants should be placed on a board or in a shallow dish, which must be frequently scalded and kept perfectly clean. Feed only what they will clean up in fifteen minutes, removing what is left, as the poults must not have sour food. If the food that is left is not removed, you invite bowel trouble and this is difficult to overcome. At all times water and fine grit should be kept before them. The water must be clean and not be left standing in the sun.

"The first food may consist of boiled fresh eggs, or custard made of one egg to a tablespoonful of milk, baked dry. The first day, use either the boiled egg or custard; after that mix the egg or custard with a good grade of commercial growing mash, or fresh raw eggs mixed with the growing mash, thoroughly scalded. When using scalded food, allow sufficient time for it to cool and to absorb the flavour of the different ingredients; it should be moist, but not wet or sloppy, nor should it be allowed to become sour and ferment. Gradually reduce the ration of egg or custard until at the end of the second week they have mash only. Continue with the mash, either moist or dry, gradually adding small grains, such as chick grain, millet or canary seed. The wet mash can be mixed with either water or milk of any kind. When using curd, see that it is free from alum, as this drug interferes with digestion. At all times the young must have finely cut, succulent greens, such as lettuce, clover, alfalfa or chick weed; give all they will eat. As the poults mature, use larger grain until they are on the regular adult ration.

"When poults are not growing properly, increase the volume of milk either in the form of fresh milk to drink or curd fed in a small dish. If maturing too rapidly, reduce mash and increase grain, thus reducing amount of protein."

After this extensive description of the methods of feeding, it may be appropriate to include some information about the housing of young pheasants.

When the eggs have been hatched by a pheasant hen or a hen the poults are accommodated in a small coop which is very simple to make. This run can be constructed of a few thin pieces of wood and some wire-mesh netting, and should be simple to build even to the most inexperienced amateur carpenter. There are no fixed measurements, but it is advisable to make the run neither too small nor too large. To give some idea of a practical size I would suggest the following measurements: width $2\frac{1}{2}$ feet, length $3\frac{1}{2}$ feet and an approximate height of just under 2 feet. The floor can be made quite well out of wire-mesh or of wood if preferred. It should be kept in mind that this run should offer complete protection against wind and rain, so a glass cover over at least part of the run is required. The remaining part of the roof could be formed by a movable hinged lid to facilitate feeding the chicks.

In addition, part of the run should be screened off by wide wire-mesh netting (see Fig. 5) through which the chicks can easily pass but preventing the hen from entering the second part of the run. The screened-off part is used for feeding the chicks their special mash and permits them a refuge if the (foster-)mother should become too overbearing.

If there is a sufficiently large stretch of turf it is highly recommended to move the run to a fresh site every day, hence the wire-mesh floor. If this is impossible the coop should be provided with a thickly spread, fresh layer of sand every day, on a wooden floor. The run should never be placed on the same stretch of grass where shortly before another run with pheasant chicks has been, as this would expose the second batch to infection.

The outside run should, of course, be connected to a night coop that is easily closed, which facilitates changing the position of the run.

All this data should be regarded as giving general instructions only. Special varieties need special treatment, to which attention is given in the relevant chapters.

Crooked Toes

The importance of correctly feeding the young pheasant chicks right from the beginning will be evident from the following. Every pheasant breeder is familiar with the evil of crooked toes. This malformation occurs between the ages of a few days and of about six weeks. It can be limited to one toe, but all toes on one or even both legs may be affected. The trouble may become serious to such a degree as to make it almost impossible for the bird to walk. Of course, such malformations (even if only one toe is affected) vastly impair the beauty and value of a bird.

In the old days it was assumed that the trouble was due to intensive inbreeding. It has since been established that the malformation is caused by incorrect feeding, not only of the chicks but of the parents as well.

The lack of certain minerals in the food combined with cramped quarters are the main causes of the, unfortunately quite frequent, malformation. Faulty artificial incubation is also cited as a cause of crooked toes, if the temperature inside the incubator has been too high.

It is possible, though very difficult, to remedy the malformation. If the entire feet are affected correction is out of the question; but if the trouble is limited to one or just a few toes a diet containing phosphates and vitamins may prove beneficial.

Another trouble pheasant-breeders have to contend with is feather-pecking. The pecking of each other's feathers among the chicks may become so serious that the birds die, robbed of practically all of their plumage. This is also caused by faulty care. Lack of space and a shortage of proteins or of cellulose are usually to blame. When these causes have been removed the feather-pecking may cease, though it can be quite difficult to break this bad habit once it has begun. The ultimate remedy is to file or burn off part of the upper bill or the fixing of so-called "spectacles", though the later is both impractical and cumbersome if there are a large number of birds.

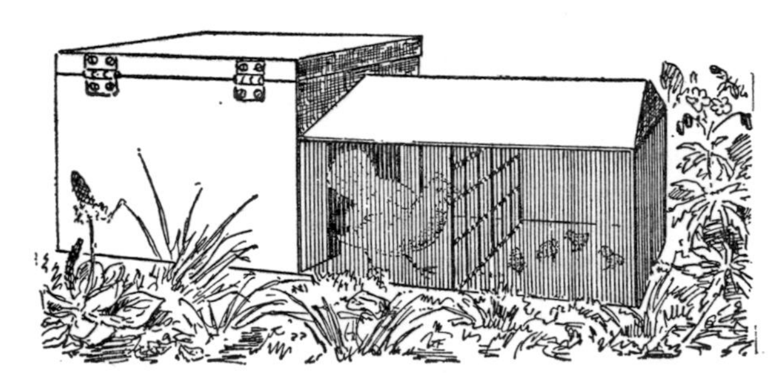

FIG. 5. Suggestion for a coop, which allows a separate compartment for the chicks.

The Ruffed Pheasants

(*Chrysolophus*)

THE Ruffed Pheasants—probably the most beautiful and popular of the entire pheasant genus—include only two species, which although closely connected, differ greatly in pattern and colours. They are the Golden Pheasant (*Chrysolophus pictus*) and the Lady Amherst Pheasant (*Chrysolophus amherstiae*). These two species interbreed freely. All hybrids are completely fertile, and endless combinations have been produced. Some beautiful specimens have been bred by these crosses, but, on the whole, the results have been rather disappointing, and—though this is a matter of taste—the splendour of the original parents has never been matched, let alone surpassed. Hybrids can also be obtained with all species of Phasianus (True Pheasants), Syrmaticus (Long-tailed Pheasants) and Lophura (Gallopheasants). But only the males are fertile; all hens prove to be sterile and are often masculine in plumage. It is also possible to cross Chrysolophus with Lophophorus (Monals) or Gallus (Junglefowls); these hybrids are, however, all sterile.

All male Ruffed Pheasants are adorned with a broad, silky crest and a large ruff of long, broad feathers which can be spread like a fan around the head and neck. Most of the other body feathers are highly specialized; that is to say, they are either disintegrated and silky or broad and scaly.

The tail is very long, composed of eighteen rectrices, graduated and ending in a point, formed by the two middle feathers, which are longest. The wings are short and rounded, the legs long and slim with a short spur. The bill is slender.

Ruffed Pheasant males come to full splendour in their second year. They have a juvenile plumage in the first year.

The females are barred dark brown and buff, with almost plain abdomens. Their crests and ruffs are greatly reduced and show the same colour as the other feathers. The face, in contrast to the males, is scantily feathered round the eyes, so that the skin is clearly visible.

Immature birds resemble the female, but are slightly lighter in colour and less strongly barred. After the first moult young cocks are easily distinguished from the hens. The young cocks have a special plumage in their first year; this dress appears when they have moulted for the first time, and they are then easily recognized; head, neck and shoulders show distinctly coloured feathers.

The eggs number five to twelve in a clutch, though annuated hens, especially in an aviary, often produce many more. The eggs are light brown in colour and have no spots.

The genus Chrysolophus (i.e. Golden and Lady Amherst Pheasant) was originally found on the mountains of Central and Western China, and a few representatives of the species have been found in Tibet and Burma.

To avoid difficulty in distinguishing between the hens of the two species it is well to bear in mind that the most striking differences are that the Golden Pheasants have *buff-coloured* legs and orbital skin; the Lady Amherst female shows bluish-grey skin in these places.

In a natural state the Ruffed Pheasants inhabit rocky, overgrown mountain-sides, avoiding the forests as much as possible. They go mainly in pairs or small parties. They are wholly indifferent to cold and feel perfectly at home both in Europe and America.

It is difficult to find a better aviary bird. Ruffed Pheasants are easy to keep and to rear; they breed well, and each cock can be housed without difficulty with two or three hens. Their size is small for a pheasant, and this, combined with their tameness and dazzling beauty, has made them a long-standing favourite of the pheasant fancier. Over the years they have become so well acclimatized and have adapted themselves so thoroughly to confinement that they are probably more common today in aviaries than they have ever been in the wild state.

The chicks are small but strong and easily reared; it makes no difference whether this is done by their own mother, by a foster-mother (e.g. a Bantam) or in an artificial incubator.

Although the cocks fight, it is often possible to keep a large flight together in one aviary, so long as no hens are near.

These pheasants can very well be kept at liberty, but it must be remembered that each pair, or group of cock and several hens, will appropriate a piece of ground for themselves and chase away all intruders. Even their own young are soon considered unwanted guests, so they will quickly set out to find another suitable area.

Ruffed Pheasants are fortunately not used as game-birds. When approached they do not fly up, as do Wood Pheasants, but they will try to save themselves by running away at great speed and by skulking.

The courtship of Ruffed Pheasants is always a wonderful performance. The crest is raised and the ruff is spread like a fan, so that it completely hides one side of the neck and part of the head. While doing this the cock continually walks round the hen, all the time spreading that side of the ruff which is turned to her. A peculiar whistling sound accompanies the courtship; the back of the bird is shown to the best advantage when the tail is opened vertically displaying its splendid colouring.

It has been mentioned that the males of the Ruffed Pheasant are fully mature only in their second year, and that hence they do not assume their full dress until then. This does not prevent many cocks, as well as hens, raised in captivity being ready for breeding in their first year. Breeding with such young pheasants, however, is not recommended; also, better results are obtained with annuated birds.

THE GOLDEN PHEASANT (*Chrysolophus pictus*), Plate V

It is difficult to describe fully a bird so abundantly coloured as a Golden Pheasant cock. The following description has been kept as concise as possible; it does not claim to be complete. It is hoped, however, that readers will get some idea of the appearance of this variety and the renown it undoubtedly enjoys.

COCK. On the crown of the head is a crest of elongated, silky, bright golden-yellow feathers. The ear-band is a brownish grey, the other parts of the face, the chin, the throat and the neck are brownish red. The tippet is formed of broad, rectangular feathers, the visible part of which is light orange in colour. Every feather in the tippet has two dark-blue bars across the tip. The upper part of the back is deep green, and every feather is margined with velvet black. The lower part of the back and the rump are of a deep golden-yellow colour. The tail-feathers are mottled predominantly black and brown. The wings feature the colours dark red on the wing-tips; deep blue tertiaries; black and brown bands on the primary and secondary quills. The entire under-part of the bird is scarlet, merging into a light chestnut in the middle of the abdomen and the thigh-parts. Under-tail coverts are red. The iris and the naked skin round the eye are light yellow; the bill and legs are a horny yellow.

HEN. The plumage of the hen is much plainer than that of the cock. The colours are mainly light, medium and a very dark brown, with an occasional pale-yellow feather. The feathers show a black mottled or barred design. It should be noted in passing that among the hens variations in the intensity of colour do occur.

The iris is brown, the skin round the eyes is yellow, the bill and the legs are a horny yellow.

For comparison, here are the average sizes of the cock and the hen:

COCK: length 1,000-1,100 mm (40–44 in.); wing 190–220 mm ($7\frac{1}{2}$–9 in.); tail 775–790 mm (31–31·6 in.); tarsus 75 mm (3 in.).

HEN: length 640–670 mm ($25\frac{1}{2}$–27 in.); wing 165–180 mm ($6\frac{1}{2}$–7 in.); tail 350–375 mm (14–15 in.); tarsus 70 mm (2·8 in.).

The Golden Pheasant originates from the mountains of central China; the exact boundaries of its entire native country are not known.

In general, it is striking how little knowledge of the life of the Golden Pheasant in its native habitat exists. This is all the more amazing when we consider the enormous popularity of this splendid bird; especially as in China it has played for centuries an important part in art and literature. Nowhere can be found even a slight description of the way in which the bird, in its wild state, makes its nest.

In the eighteenth century the Golden Pheasant was already quite well known in Europe. This shows how extremely durable the popularity of this pheasant has proved to be. It is, *par excellence*, an aviary bird. Even if the aviary be relatively small, the Golden Pheasant will feel completely at home.

It is not bothered by the cold and will easily endure the most severe winter. The Golden Pheasant is a very good breeder. Its eggs are found in early spring, and, what is quite as important, they are often laid in large numbers, varying from ten to twelve eggs a season for young hens, to between thirty and forty for hens more than a year old.

The Golden Pheasant hen is a faithful broody and rarely leaves the nest for food and drink during the entire brooding period. This covers twenty-two days. As a result of numerous cross breedings with the Lady Amherst Pheasant many Golden Pheasants have lost their purity of race. True Golden Pheasants are slender and long-legged. The face and throat are of a uniform red, and the tail is abundantly covered with black spots, *without any trace of a barred design*.

According to Delacour, the Golden Pheasant cock is very sensitive to sunlight, which discolours the golden-yellow parts of the plumage as well as the orange tippet. After the moulting the original colours return, but quickly lose their brightness. To prevent this it is advisable to keep the Golden Pheasant in the shade as much as possible during the summer.

There also exists two mutations of the familiar Golden Pheasant (*Chrysolophus pictus*). One is the so-called Dark Golden Pheasant (*Chrysolophus pictus obscurus*). This bird, too, is of gorgeous appearance, but, as a rule, is rather rare. At the moment, however, many breeding pairs are available, though mostly in America.

The general appearance seems at first to resemble the ordinary Golden Pheasant, but the bird is darker in colour. The greyish-black shade of the face, the throat and the upper part of the breast are the most obvious characteristics. There should be no trace at all of a metallic-green lustre. If this should be found, then the bird is not the rare dark variety but an "ordinary" hybrid of the normal Gold and the Lady Amherst.

The hens, too, of this rare Dark Golden Pheasant have a much darker hue than those of the normal variety. The basic colour of the former is a dark reddish brown, with the upper part of the throat and the underside of the body of the same deep colour as the rest of the plumage.

The second is the fairly recent yellow mutation, (*Chrysolophus pictus luteus*). An unusual cock bird was hatched from an ordinary Golden Pheasant egg some years ago in Germany, and fortunately it was seen by a German fancier, who arranged for Professor Ghigi of the Bologna University to acquire the bird. This cock was mated with a normal hen and produced only normal young. The hereditary colour factor was, therefore, recessive, but pairs of this off-spring produced 50 per cent of the new mutation. From this bird, Professor Ghigi produced the present-day Yellow Golden Pheasant, which is entirely the result of a mutation and not of cross-breeding. (See photo Fig. 19, p. 115).

Apart from these Golden Pheasants, no other varieties or mutations exist. All other Golden Pheasants that feature a divergent pattern of colour or plumage are a result of cross-breeding (which may be selective breeding) with the Lady Amherst Pheasant. No doubt such hybrids may have a splendid design of colours, but they are of no special value.

THE LADY AMHERST PHEASANT (*Chrysolophus amherstiae*), Plate V

In order to give an idea of this gorgeous bird, a short description of the colouring of the Amherst Pheasant is also included. Unfortunately it will be found that formerly—because of the lack of sufficient pairs for breeding—a great deal of cross-breeding with the Golden Pheasant was carried out with the result that today there are many impure Amherst Pheasants.

To enable the pheasant fancier to distinguish between pure and impure a rather more elaborate description of the colours of the birds than has been given hitherto is included. Delacour's wonderful description of the Amherst Pheasant seems most suitable for this purpose, so it is here given in full:

COCK. Crown covered with short metallic-green feathers; narrow nuchal crest of stiff, elongated crimson feathers; ruff of rounded feathers, white with a blue-and-black border; mantle and scapulars of rounded feathers, metallic bluish-green with a black border edged with scintillant green; feathers of back broad and square, black with a green bar and a wide, buffy yellow fringe; those of the rump with a vermilion fringe; tail-coverts mottled black and white with long orange-vermilion tips; central rectrices irregularly lined black and white with black cross-bars; other rectrices similar on the narrow inner web, silver-grey passing to brown outside, with curved black bars on the outer web; wings dark metallic blue with black borders, the primaries only blackish brown sparsely barred with buff; face and throat black, with metallic-green spots, breast like the mantle, the borders of the feathers wider and brighter; rest of the under-parts pure white, the base of the feathers grey, except the lower flanks and vent, which are barred with black and brownish grey; thighs mottled white, black and brown; under-tail-coverts black and dark green more or less barred with white. Iris pale yellow; bare facial skin and lappet bluish or greenish white; bill and feet bluish grey.

HEN. Similar to the Golden Pheasant hen, but larger, the dark barring blacker, with a green sheen; crown, sides of head, neck, mantle, lower throat and upper breast strongly washed with reddish chestnut; upper throat and abdomen pale, sometimes white; lores, cheeks and ear-coverts silvery grey spotted with black; back strongly vermiculated; tail-feathers rounded, not pointed at the tip, as in the Golden Pheasants, and much more strongly marked with broad irregular bars of black, buff and pale grey vermiculated with black. Iris brown, sometimes pale yellow or greyish in older birds; orbital skin light slaty blue; bill and legs bluish grey.

For comparison of the sizes of male and female the average figures are given below.

COCK: length 1,300–1,700 mm (52–68 in.); wing 205–235 mm (8·2–9·4 in.); tail 806–1,150 mm (32·3–46 in.); culmen 25 mm (1 in.); tarsus 75–85 mm (3–3·4 in.).

HEN: length 660–680 mm (26·4–27·2 in.); wing 183–203 mm (7·3–8·2 in.); tail 310–375 mm (12·4–15 in.); culmen 23 mm (1 in.); tarsus 66–74 mm (2·6–3 in.).

Plate III

Malay Crested Fireback or Vieillot's
Lophura ignata rufa

Siamese Fireback
Lophura diardi

Swinhoe's Pheasant
Lophura swinhoei

Brown Eared Pheasant
Crossoptilon mantchuricum

The Lady Amherst Pheasant is found in a wild state in the south-east of Tibet and south-western China. This species lives in a more southerly region than does the Golden Pheasant, but, on the other hand, in much higher places, which explains why it is just as hardy. But the knowledge of the habits and way of life of the wild Amherst Pheasant is as scanty as the information about the Golden Pheasant. All that is now known is the approximate area of distribution and the fact that these birds mostly live in mountainous regions. It has also been discovered that the Lady Amhersts unite into groups of twenty to thirty at the approach of autumn and winter—as opposed to the Golden Pheasant—and that when pursued they break into flight sooner than do the latter.

It was Lord Amherst who first introduced into Europe this pheasant in a live state—in the year 1828—by sending two males to London. The birds did not live long, however, and it was not until 1869 that a few specimens were again brought here.

After some more ups and downs the number of Amherst Pheasants in Europe steadily increased by the arrival of a few more specimens and by successful breeding. The Lady Amherst adapted itself very quickly to life in an aviary and soon proved to be just as easily kept and bred in confinement as the Golden Pheasant. The chicks may be a little more sensitive, and the cocks somewhat more aggressive, but by and large the Lady Amherst has proved an ideal aviary bird.

After the arrival of more specimens in the 'seventies no imported Lady Amherst Pheasants were available for a long time. It may therefore be safely assumed that all Lady Amhersts in America or elsewhere have originated from specimens bred in Europe.

It is, however, a pity that at first more males were imported than females. The result was that the remaining cocks were cross-bred with Golden Pheasant hens. This continuous practice of reverse breeding has produced a large number of hybrid Amherst Pheasants, and a pure Lady Amherst is today very rare.

A Golden Pheasant strain in a Lady Amherst Pheasant shows itself (among other characteristics) in a greater or lesser degree of red colour on the flanks and thighs. Traces of a mottled design on the tail, scarlet instead of orange tips of the tail-coverts and too big a crest, are all characteristics of the impurity of the variety.

Occasionally a Lady Amherst Pheasant cock may be found of an entirely pure design with the crown of the head wholly scarlet in colour. This is also a sign of cross-breeding with the Golden Pheasant, and it is advisable not to use such birds for further breeding.

Females may be found without the deep red hue on the head and the upper and lower part of the throat—a deficiency that points to the presence of Golden Pheasant blood.

In general, it may be said that the Lady Amherst Pheasant is easy to breed. The hens make good broodies and the number of eggs laid usually amount to five to twelve for young birds and considerably more for hens over a year old.

The incubation period lasts for twenty-three days, and the devotion of the mother to her poults is very great.

The Long-Tailed Pheasants
(*Syrmaticus*)

THE genus of the Long-tailed Pheasants consists of five different species, the representatives of which are easily distinguished from each other. Two are closely related, making it rather more difficult for the layman to tell them apart. But once one has seen a coloured picture of both birds recognition will no longer be a problem.

The five representatives of the species Syrmaticus are: the Elliot's, the Hume's, the Mikado, the Copper and the Reeves's Pheasant. The Elliot's and the Hume's Pheasants are the two related species just mentioned. In addition, there are a few beautiful and pure-breeding variations of the five main races, but it is unnecessary to describe them here. After all, this book is meant not to be a complete guide to all races of pheasants but a help to the fancier.

The nearest relatives of these species are the True or Game Pheasants; representatives of the genus Phasianus. Both varieties show sufficient points of difference to be easily distinguishable. For example, the tail-feathers of the True Pheasants are much shorter, and the long-tailed pheasants do not have the tufts of feathers near the ears.

The specimens of the Syrmaticus have no crest, nor do they have a ruff, which is a characteristic of the Ruffed Pheasants.

In general, the Long-tailed Pheasant is a quiet bird which produces only occasionally its typical, rather shrill call-tone.

The display of the cocks is remarkable. They stretch their necks and at the same time expand the feathers, giving the whole a greatly swollen and puffed-up aspect. Simultaneously they rustle their wings and spread their tails horizontally. The Mikado and the Elliot's Pheasant also manage every now and then to display the tail, peacock-fashion, but this is the absolute limit of their powers of display.

In contrast to the Ruffed Pheasants, all representatives of the genus Long-tailed Pheasant assume their full dress in the first year.

The hens, as a rule, resemble each other somewhat, and have a fairly intricate pattern of colours.

Young birds look like the hens, but are slightly dully coloured. The number of eggs in a clutch varies between five and fifteen.

The Long-tailed Pheasants inhabit mainly the wooded mountain slopes of Burma, Siam, Formosa, China and Japan. The heights at which they are found vary with the different species. In a wild state they largely feed on fruits and insects. They inhabit regions with a temperate climate and have proved completely adaptable to the heat of the sun and the cold of winter, both in Europe and large parts of North America.

Their magnificent plumage, their easy upkeep and their excellent breeding qualities have given them a good reputation as aviary birds. The only objection to them is their combativeness and their fierce temperament, although these bad qualities are more pronounced in the one species than in the other. The Copper or Soemmering's Pheasant is the most malevolent, the Mikado the most manageable, among the Long-tailed Pheasants.

Three of the five species have completely adapted themselves to confinement, viz. the Reeves's, the Elliot's and the Mikado. The Copper Pheasant has been kept in aviaries at various times, but unfortunately has never proved a great success. This is due to the bad habit of the male in sooner or later killing his hens.

Hitherto the Hume's Pheasant has never been imported into Europe or America, so neither this variety nor the Copper Pheasant need be discussed here.

The Reeves's Pheasant is nowadays common in European pheasantries and has also been successfully introduced as a game-bird.

The Elliot's Pheasant is quite numerous in America, less in Europe. This pheasant is easily bred, and it will therefore probably return in due course to its pre-war popularity in Europe as well.

The Mikado Pheasant is rare, but a few breeding couples exist in Europe and the number may be expected to increase.

It has already been mentioned that the birds of the genus Long-tailed Pheasant assume their full dress, and consequently are ready for breeding, in the first year. The fact remains, however, that hens more than a year old lay a far greater number of eggs; this may be an indication of inbreeding or specific shortcomings in the care of the birds, lack of space or other causes.

The combative nature of the males compels the pheasant-breeder to supply ample shelter for the hens in the aviary. It is even a common practice to clip the wings of the cocks, thus making it impossible for them to pursue the hens to the higher perches. It is also always advisable to house the cocks and the hens separately when the breeding period is over.

The chicks need almost the same treatment as those of the other pheasants. As a rule, they are not very susceptible to contagious diseases.

Elliot's chicks as well as Copper's are tiny and rather fragile, especially soon after being hatched. Feeding live insects during the first two weeks is particularly recommended; it is difficult to induce the chicks to eat unless something moving attracts their attention.

The various species of the Long-tailed Pheasants can easily be cross-bred, though the resulting hybrids are only rarely fertile. As a rule, all hens are infertile. This is also the case when Long-tailed Pheasants are crossed with Gallopheasants, Ruffed Pheasants and True Pheasants.

THE ELLIOT'S PHEASANT (*Syrmaticus ellioti*), Plate V

Before the war this variety was quite common, but nowadays it is unfortunately very seldom seen. It is gratifying, however, to note that a few good breeding couples are now available, so the numbers of these beautiful birds may be expected to increase rapidly.

The purchase of a pair of Elliot's will be beyond the means of most breeders. Furthermore, this pheasant is not a bird for beginners, notwithstanding its good breeding qualities. It would be a pity to trust it to an inexperienced owner, and a breeder of long standing knows enough to recognize the species and to understand its peculiarities.

The Elliot's Pheasant inhabits a limited region of China, called Chekiang, where it was discovered in 1872 by Swinhoe. Shortly afterwards these pheasants were introduced and successfully bred in Europe, when the bird proved to be reasonably fertile and wholly adaptable to our inclement climate. At present it is rarely found in China because of a progressive cultivation of its restricted native region and also by reason of much persecution. Fortunately there are many Elliot's Pheasants in aviaries in America, and there has been a noticeable increase in Europe.

In order to prevent serious fights between the male and the female (such a fight usually ends in the death of the hen) it is advisable to provide two or three hens to a cock; moreover, it is necessary to plant the aviary with thick shrubbery in several places to provide adequate shelter for the hens.

The Elliot's hen is a moderately good brooder, laying a clutch of ten to twenty eggs, which are hatched after twenty-five days.

The poults are small and wary and require much attention during their first days in order to induce them to eat. Fresh ants' eggs, grubs and mealworms yield good results. After a few days the most difficult period is over and the chicks are easy to rear. The young males may be distinguished at an early stage from the females because of the barred middle feathers of the tail.

The following are the measurements of the full-grown Elliot's Pheasant:

COCK: length 800 mm (32 in.); wing 230–240 mm (9·2-9·6 in.); tail 390–440 mm (15·6–17·6 in.); culmen 30 mm (1·2 in.); tarsus 70 mm (2·8 in.).

HEN: length 500 mm (20 in.); wing 210–225 mm (8·4–9 in.); tail 170–195 mm (6·8–7·8 in.); culmen 27 mm (1 in.); tarsus 63 mm (2·5 in.).

THE MIKADO PHEASANT (*Syrmaticus mikado*)

The male of the Mikado Pheasant is predominantly blue-black in colour, with a white-barred design on the tail and wing-coverts. The face around the eye is a bright scarlet and the head and neck have a bluish-purple lustre. The female is of a rather dark olive brown, with a white design on the coverts. The middle feathers of the tail are chestnut with large black dots.

The sizes of the Mikado Pheasant are:

COCK: length 875 mm (35 in.); wing 210–230 mm (8·4–9·2 in.); tail 490–530 mm (19·6–21·2 in.); culmen 26 mm (1 in.); tarsus 67 mm (2·6 in.).

HEN: length 528 mm (21·2 in.); wing 187–215 mm (7·5-8·8 in.); tail 172–225 mm (6·8–9 in.); culmen 24 mm (1 in.); tarsus 57–61 mm (2·2–2·5 in.).

The Mikado Pheasant is found exclusively in the mountainous areas of Formosa, at heights varying between 6,500 and 8,000 feet. Almost without exception these birds inhabit dense forests of oaks and bamboos. They are found only at high altitudes. In a wild state their food consists of berries, seeds, greenstuff and insects; more or less the same diet of the other species of Long-tailed Pheasants.

The Mikado has been known for only a short time. It was first heard of in 1906, and it was not until 1912 that the first pairs were imported into Europe. The Mikado was bred for the first time in 1913 and 1914. It was found that successful rearing needed a great deal of green food. Experience also showed that the Mikado Pheasant is hardy and robust, though susceptible to foggy or damp winter weather.

FIG. 6. Mikado Pheasant
(*Syrmaticus mikado*)

Quite a few hybrids between a Mikado cock and an Elliot's hen have been reared.

The normal laying period is from the end of February to the beginning of May; the hens lay an average of five to ten eggs, and twenty-six to twenty-eight days are needed for hatching.

Enough Mikado Pheasants are now available in Europe and in America to ensure the survival of the species, and gradually to increase the number of birds. All the same, this pheasant should still be considered a rare bird and it is a far from common sight in a pheasantry.

While the Mikado Pheasant might appear to be dark and dully coloured, it is, in reality, very beautiful. This species is hardy to our climate, and is easy to keep and to rear. The hens, if well fed, will lay in their first year, but here, too, it is always preferable to breed with two-year-old birds.

In the aviary the hen of the Mikado Pheasant can lay three clutches of eggs between the end of March and the middle of June, sometimes as many as thirty eggs in all, usually fifteen to twenty.

As a rule the chicks are quite easy to rear, as they are naturally tame. They are, however, very susceptible to infections, as are all pheasants living at high altitudes. The thin air of their native regions contains far less bacteria than does the air at the low altitudes into which the birds are introduced.

These birds consequently have a low resistance to all kinds of diseases. The right diet and a high degree of cleanliness give very good results, and the breeder who is prepared to go to extra trouble over his chicks is sure to have healthy and strong birds.

The Mikado cock, like all Long-tailed Pheasants, is rather aggressive towards the hens, but when adequate precautions are taken—such as planting the aviary—two, and even three, hens can be kept with one cock.

The Reeves's Pheasant (*Syrmaticus reevesi*), Plate V

A description of the colours of this beautiful and extremely well-known pheasant would seem to be superfluous. Nearly every pheasant fancier will have a couple of Reeves's Pheasants in the aviary, and everyone will have seen a Reeves's Pheasant, either in the flesh or on a coloured print.

From the measurements given below it will be realized why the genus "Syrmaticus" is called "Long-tailed Pheasant".

Cock: length 2,100 mm (84 in.); wing 275–300 mm (11–12 in.); tail 1,000–1,600 mm (40–64 in.); culmen 34 mm (1·3 in.); tarsus 80 mm (3·2 in.).

HEN: length 750 mm (30 in.); wing 275–300 mm (11–12 in.); tail 360–450 mm (14·4–18 in.); culmen 28–32 mm (1·1–1·3 in.); tarsus 62–65 mm (2·5–2·6 in.).

The Reeves's Pheasant is also a native of China and is found in almost all high wooded areas in Central China. Though the birds are at present still quite common in China, their number is visibly diminishing. This gradual decrease in number is undoubtedly caused by the growing cultivation of the areas where they live. The same principle applies to almost all birds and beasts living in a wild state, in Europe as everywhere else.

The Reeves's Pheasant is found at altitudes varying between 750 feet and, in some cases, 2,000 feet. Its way of life and habits very much resemble those of the other representatives of the species Syrmaticus.

The Reeves's Pheasant is a very fast and expert flier. Its long tail, which might be considered quite a handicap by the casual observer, by spreading, is used as a kind of brake, allowing the bird to turn and stop abruptly.

The display of the cock is simple. It is restricted to puffing out the feathers, mainly those of the neck and the tail. Before his excitement has reached a peak he walks proudly up and down, without, however, paying the slightest attention to the hen. Only at the moment of greatest excitement he jumps towards the hen, pressing his head against his back.

The first Reeves's Pheasant—a cock—came to Europe in 1831, but not until 1867 were the first breeding results obtained. As usual, it was tried to obtain hybrids with the Reeves's cock, because of a lack of hens. Hybrids were produced by mating a Reeves's cock with a Common Pheasant hen, and one of the hybrid hens proved fertile.

Once the Reeves's Pheasant had been bred in Europe its number increased quickly and soon became very great. The birds were successfully introduced into woods and shooting preserves, and today great numbers of wild Reeves's Pheasants are found in wooded areas both in England and France. The birds did not prove very successful game-birds. They are very reluctant to leave cover, making it next to impossible to shoot them. Another unpleasant characteristic of the Reeves's Pheasants is their extreme combativeness, driving other pheasants away from their region.

It is evident from the excellent breeding results obtained with Reeves's Pheasants that this bird is completely hardy in the European climate. Its magnificent plumage, its easy care and its very good breeding properties make this pheasant one of the most popular aviary birds. A drawback when keeping Reeves's Pheasants in the aviary is their excessively long tail. To avoid its being damaged a spacious aviary is required.

It is preferable to keep two or more hens with one cock, as the male Reeves's is extremely pugnacious. If enough shelter is provided in the aviary by planting shrubs and small trees the fights will not be of great consequence. If the cock continues to interfere with the hens it is better to take him out of the aviary after mating.

The chicks are fairly small, but they are strong and easy to rear. Even with quite small chicks attention should be paid to their hot temper, and breeders would do well to avoid rearing Reeves's chicks in mixed heads. If they are raised together with other breeds there will certainly be victims. Even among themselves the young poults soon begin to fight. The only remedy is a very large run.

A clutch of eggs of the Reeves's hen may consist of some twenty eggs. Several clutches produced in one season is quite normal. The incubation period is twenty-four to twenty-five days.

Apart from hybrids obtained with the Common Pheasant, Reeves's Pheasants have been successfully mated with Copper, Cheer and Silver Pheasants. Only the hybrid cocks are fertile.

The Gallopheasants
(Lophura)

A NUMBER of pheasants are united in the genus "Gallopheasants"—to be exact, ten species and a few sub-species—which resemble one another generally in build, shape and life habits. This was not always the case, and the genus Lophura was subdivided into a few separate genera. Research during the last few years has clearly shown that all ten species are related to a sufficient degree to form one single genus. Hence there are to be found in the genus Lophura some rather primitive forms and also some highly specialized pheasants. They are heavy, fowl-like birds with compressed, half-bent tails and large, erectile wattles covering the face round the eye. The legs are heavy, and each leg in the male is armed with a long, sharp, white spur. Sometimes the hen has spurs as well. The bill is moderately short, but strong. The difference between the two sexes is striking. The cocks, as a rule, show various tones of grey, chestnut, blue, black or white, with the addition of crimson, yellow and coppery red in some species. The hens, on the other hand, show all kinds of shades of brown with or without markings or stripes. The one exception is the hen of the *Lophura Erythrophthalma* (one of the Fire-backed Pheasants), which closely resembles the male counterpart.

Among the ten main species of the Gallopheasants three are fully crested in both sexes, in four the cock alone possesses a crest, while three are entirely crestless. Shape, size and structure of the crests in the various species are conspicuously different.

The face-colour is red in eight species and blue in two.

The Gallopheasants are attractive, but sometimes they are rather heavy birds. Even those which do not show bright colours—e.g. the Nepal Kalij and the Silver Pheasant—are beautifully marked and have a graceful form. The other species are outstandingly lovely. In a wild state they prefer to live in woodlands, bamboo forests and heavy brushland. They are neither wild nor shy in disposition.

In general, there is no complete picture of the life-habits of Gallopheasants in a natural state. For example, it cannot be said with any certainty whether they are polygamous or monogamous, but considering the usual habits of pheasants, it may be assumed that they are naturally polygamous. After all, mating for life or, at least, to one definite partner, as is seen in geese, has up till now not been conclusively proved of any species of pheasants.

The Gallopheasants live and nest on the ground, roosting in trees only at night. The cocks are quarrelsome, and when kept in an aviary may attack their keeper.

When Gallopheasants are kept in a pheasantry one soon notices the habit of the males of beating their wings at a fast speed. This is done at such a rate that a whirring noise is produced. With one exception, the Gallopheasants have a fairly simple courtship. They spread the tail-feathers vertically, expanding the crest and the face wattles and walk round the hens, shaking the tail.

The Gallopheasants can be kept in a fairly small aviary and have very simple food requirements. Consequently they are very easily kept in captivity. They are as a rule very tame, even when kept in pairs, and are easily induced to breed. All this has made them very popular aviary birds that are found everywhere. Of course, not all pheasants belonging to the species are found in great numbers, and there are some representatives of Gallopheasant which are not found in captivity at all. The above only applies to the well-known varieties, according to the purpose of this book.

With the exception of the Firebacks, all Gallopheasants are hardy. The hens do not build a proper nest, they find a sheltered spot in which they deposit their eggs. The size of the clutch varies a great deal. The incubation period lasts twenty-one to twenty-five days, depending on the species. An average of twenty-four days can be expected with the most common varieties.

Although all pheasants hitherto described have been distinguished with a single addition after the name of the genus, it will not be possible to maintain this principle when naming the species of the genus Lophura. Some of the pheasants are so closely related and often look so much alike that a double name does not provide sufficient distinction. The reader will notice that some pheasants now being discussed are all classified as Lophura Leucomelana and only the third qualification, such as "hamiltoni", "melanota" or "lathami", makes it possible to indicate accurately the variety in question.

THE WHITE-CRESTED KALIJ (*Lophura leucomelana hamiltoni*)

This pheasant was originally imported in fairly large numbers. Of late, however, the number of White-crested Kalijs has greatly diminished, due to hybridizing and to a decline in interest.

From the description which follows—and to which reference will be made when discussing some of the other species of this genus—it will be evident why the interest in this pheasant is decreasing. It will soon be realized that the White-crested Kalij is a handsome bird but not at all a spectacular one. The competition of other pheasants, more attractive to look at, has proved—unjustly so—too great.

From the following description it will be possible to form a picture of the White-crested Kalij.

COCK. At first sight the bird makes rather a dark impression, with the colours blue and black by far the most important. When studying the bird at close range, however, quite a few different shades of colour will be noted. The crest, which appears to consist of hairs rather than of feathers, has a white colour (in some cases a dull shade of brown grey). The head, the neck, the mantle, the back, the rump and the lesser tail-coverts are all purplish or bluish-black; the feathers of the mantle have pale grey edges and white shafts. The feathers of the back, rump and lesser tail-coverts have a brown line and a broad white terminal fringe. The tail is entirely black, but for brown tips of the feathers. The black wings have brown reflections, and this colour pattern is repeated in the throat and chin. The breast has striped feathers, brownish grey at the base and light grey to

FIG. 7. White-crested Kalij.
(*Lophura leucomelana hamiltoni*)

whitish at the tip. It should be kept in mind that the colour pattern of the crest and on the breast of this pheasant may show differences in various birds of the same species. The feathers of the flanks and the abdomen are wider and are dark grey with white shafts and pale edges. The bill is greenish white, with a dark base. The legs are brown grey and the colour of the iris varies between a yellowish orange and brown. The face-wattles are a striking scarlet.

HEN. In general, this hen, too, is predominantly brown (this applies more or less to all pheasant hens), and only a close scrutiny reveals a striking wealth of colour. Certainly, at first sight, the pheasant hen appears a very uninteresting bird, but when comparing the different colour tones their number is impressive. The female White-crested Kalij has a pale brown plumage, but a close look shows that the feathers are vermiculated with blackish, that they have pale shafts and light grey borders, except for the flight-feathers that do not have these grey borders. The central rectrices are brown, with narrow black and buff lines, as opposed to the other rectrices which have a blackish-brown colour, with light tips. The hen has a crest as well; it is long and has a dull-brown colour. The irides, the bill, the legs and the face-wattles have the same colour as those of the cock, but a duller shade. The same rule given in the description of the cock applies to the hen: there may be considerable differences in colour and pattern among birds of the same species.

The sizes of both sexes are:

COCK: length 650–730 mm (26–29·2 in.); wing 225–250 mm (9–10 in.); tail 230–350 mm (9·2–14 in.); culmen 23–30 mm (1–1·2 in.); tarsus 75–80 mm (3–3·2 in.).

HEN: length 500–600 mm (20–24 in.); wing 203–215 mm (8·1–8·6 in.); tail 205–215 mm (8·2–8·6 in.); culmen 20–25 mm (0·8–1 in.); tarsus 65–70 mm (2·6–2·8 in.).

White-crested Kalijs are commonly found in the forests and thickets of the western Himalayas at heights varying from 1,200 to 11,000 feet. Their breeding period lasts from March to June, and they usually nest in the shelter of a thick bush or a tuft of grass, and always near water. They dig with the bill a great deal in search of food. The cocks are keen fighters.

White-crested Kalijs are hardy and robust and breed very easily. They need no special care, and so are just as easy to keep as the Silver Pheasant. Specimens with a pure white crest are the most beautiful, but they are uncommon. The first of these Kalijs came to Europe in 1857 and began to breed the following year.

Incubation lasts twenty-four to twenty-five days, and young cocks

Plate IV

Temminck's Tragopan
Tragopan temmincki

Malay Great Argus
Argusianus argus argus

Himalayan Monal
Lophophorus impeyanus

Satyr Tragopan
Tragopan satyra

begin to assume adult plumage in the first year. Consequently White-crested Kalijs will breed in the following spring.

A. Touchard, the well-known breeder of pheasants, has written as follows about the White-crested Kalij:

> "White-crested Kalijs are much like Silver Pheasants, having their advantages and their faults.
>
> "Less wild than the True Pheasants, less savage than the Silvers, they have not the same bad habit as the last named of eating their eggs. Misinformed people believe they breed only the second year, but I have obtained from a young pair twenty-two eggs which have produced sixteen chicks. It is rare, however, to get such fine clutches the first year, as the average is twelve to fifteen for first-year birds, and twenty-five to thirty for older pairs. But the percentage of fertile eggs is equal. A cock is enough for two hens, and I wonder what has made breeders believe that they should use only one hen because a second would be killed.
>
> "Hatching takes twenty-four days and is easy. The chicks are strong a few hours after birth and require no special care."

To the casual observer the Nepal Kalij looks very much like the White-crested Kalij. There are a few differences, however, such as the crest that is black and a little shorter, and narrower white fringes to the feathers of the lower back and the rump. The entire bird is somewhat smaller than the White-crested Kalij (dimensions—wing 204–233 mm; tail 250–305 mm).

The difference in the female Nepal Kalij consists of a darker colour of the plumage, showing up the grey borders of the feathers.

The area of distribution is Nepal, though the exact limits are imperfectly known. The Nepal Kalijs have the same habits and characteristics as are to be found in the White-crested species. These birds live in forests at altitudes from 4,000 to 10,000 feet, but they come lower down in the winter months.

The Nepal Kalij has long been common in Europe, but because of its unspectacular appearance has never been very popular. It is easy to keep and to breed, and, as a rule, is tame and confident.

THE BLACK-BACKED KALIJ (*Lophura leucomelana melanota*),

To obtain an impression of the Black-backed Kalij, it is easiest to visualize the Nepal Kalij without the white fringe on the feathers of the lower back and the rump. The tail and the crest are a little shorter and more rounded, but the general impression is the same. The white fringes of the feathers of the Nepal Kalij are a velvety-black fringe to the steel blue feathers of the Black-backed Kalij. The wing coverts sometimes have narrow white edges. (Dimensions: wing 215–240 mm (8·2–8·6 in.); tail 238–300 mm (9·5–12 in.).

The hen resembles very much the hen of the Nepal Kalij and may easily be mistaken for the other bird. They can be told apart by the intensity of the shade of brown, which is much greater in the Black-backed hen. The crest of this hen is shorter and the colour of the throat and the borders of the feathers varies between a buffy brown to almost white.

In general, the differences between the Black-backed Kalij and the other species of *Lophura leucomelana* are so small that nowadays it is doubtful whether this pheasant should be considered as belonging to a separate species. Gradually the opinion is growing that these are intermediate forms between the various subspecies.

THE BLACK-BREASTED KALIJ OR HORSFIELD'S KALIJ (*Lophura leucomelana lathami*), Plate II

The Black-breasted Kalij has also a great deal in common with the Kalijs described in the foregoing pages. Apart from a few minor points of difference such as a more upright crest, vermiculated black under-parts, short tail and longer legs, this bird has many of the same characteristics. The colour plate (11) makes a further description superfluous.

The hen has some of the same points of difference from the other Kalijs, viz. the upright crest, shorter tail, buffy brown throat and longer legs.

Size:

COCK: length 570–590 mm (22·8–23·6 in.); wing 210–240 mm (8·4–9·6 in.); tail 210–245 mm (8·4–9·8 in.); culmen 28–32 mm (1·1–1·3 in.); tarsus 76–84 mm (3–3·4 in.).
HEN: wing 205–230 mm (8·2–9·2 in.); tail 190–225 mm (7·6–9 in.).

The Black-breasted Kalij is found in the northern part of Assam and in Burma. In a wild state it cross-breeds regularly with related species (i.e. the Silver Pheasant).

Many of the hybrids have been given distinct names, but for the readers of this book these unstable intermediates are of no importance.

The Black-breasted Kalij—also called the Horsfield's Kalij—lives in the lower woodland areas and is usually not found at altitudes of more than 3,000 feet. Only the birds inhabiting Upper Burma ascend higher; they are even found at altitudes of up to 8,500 feet. It is known that in this species both parents stay with the young.

In captivity they have always been more of a rarity than the other species described here, even though they are just as easy to keep and to rear. They, too, were first imported into Europe in 1857 and were bred in the next few years. At present few Black-breasted Kalijs are found in the pheasantries.

THE LINEATED KALIJ (*Lophura leucomelana lineata*), Plate II

A short description is needed to acquaint readers with this rather different pheasant.

The individual vermiculations of the various specimens differ to a certain amount, but they are always of a finer structure than in the other Leucomelanas or the Silver Pheasant. The crest and the under-parts are black. The upper parts of the body have a delicately vermiculated pattern that produces a grey lustre. The sides also have striped feathers with a broad white shaft. Sometimes this shaft is spotted with black. Wings and tail are more coarsely vermiculated. The central rectrices are considerably lighter in colour and have buffy white tips and inner webs. The iris is yellowish brown; the bill is greenish white, darker at the base. The legs are a mixture of colours, and may best be described as a brownish blue-grey.

There is much individual variation in the size of the vermiculations, but they are always finer than in any other Leucomelana or Silver Pheasant.

The hen has a golden-brown colour and a darker crest. Neck and upper back show white V-shape shaft-markings. The rest of the feathers on the upper parts have pale grey tips. Chin and throat have a whitish appearance. The under-parts are a bright rufous to brown, with large white lanceolate centres to the feathers.

Sizes:

COCK: length 630–740 mm (25·2–29·6 in.); wing 220–260 mm (8·8–10·4 in.); tail 230–345 mm (9·2–13·8 in.); culmen 28–30 mm (1·1–1·2 in.); tarsus 75–88 mm (3–3·5 in.).

HEN: length 540–560 mm (21·6–22·4 in.); wing 203–235 mm (8·1–9·4 in.); tail 220–235 mm (8·8–9·4 in.); culmen 26–28 mm (1–1·1 in.); tarsus 72–76 mm (2·9–3 in.).

The area of diffusion covers part of Burma and north-western Siam. Lineated Kalijs mainly inhabit the bushy slopes, from which it is evident that this pheasant does not frequent the high altitudes where the species last described are to be found. Their habits are similar to those of other Kalijs.

The Lineated Kalijs first reached Europe in 1864, and young were reared soon after. Unfortunately, the crossing of this pheasant with other sub-species of the Leucomelana has been frequent, and this has made it difficult to procure pure-bred birds. The general treatment is similar to that of the other Kalijs, and this species is just as hardy and easy to rear. The incubation lasts twenty-five days.

The chief specimens of *Lophura leucomelana* have now been discussed, but four more sub-species should be included. They are:

Lophura leucomelana moffitti
Lophura leucomelana williamsi
Lophura leucomelana oatesi, and
Lophura leucomelana crawfurdi.

These birds are almost completely unknown in this part of the world.

THE SILVER PHEASANT (*Lophura nycthemera nycthemera*), Plate V

A better-known species of the Gallopheasants is the Silver Pheasant, but before describing the True Silver Pheasant it should be noted that there are many sub-species of the Silver Pheasant. All these forms, which sometimes show of great, and sometimes rather irrelevant, individual variations, are pure-breeding birds. They can, therefore, be considered not as the result of hybridizing or as unstable intermediates, but as entirely distinct species. Many of these Silver Pheasants, although known and described, have never been kept in European or American pheasantries. Some varieties have never even been exported alive from their native country. On the other hand, every now and then a pair of one species has come to Europe and has managed to survive for some time.

Should any specimens still be available in Europe or America their price will be tremendously high because of their rarity. Consequently they will be beyond the means of most pheasant-fanciers. It will therefore be unnecessary to discuss these pheasants in this book, but their names are included for the benefit of those readers who may be interested. They are:

Lewis's Silver Pheasant—*Lophura nycthemera lewisi*
Annamese Silver Pheasant—*Lophura nycthemera annamensis*
Boloven Silver Pheasant—*Lophura nycthemera engelbachi*
Bel's Silver Pheasant—*Lophura nycthemera beli*
Berlioz's Silver Pheasant—*Lophura nycthemera berliozi*
Ruby Mines Silver Pheasant—*Lophura nycthemera rufipes*
Rippon's Silver Pheasant—*Lophura nycthemera ripponi*
Jones's Silver Pheasant—*Lophura nycthemera jonesi*
Lao Silver Pheasant—*Lophura nycthemara beaulieui*
Fokien Silver Pheasant—*Lophura nycthemera fokiensis* and
Hainan Silver Pheasant—*Lophura nycthemera whiteheadi*.

A short description of the True Silver Pheasant, which is found in great numbers in pheasantries everywhere, will enable those who have not yet encountered this magnificent bird to obtain a better idea of it.

Cock. This pheasant is the largest and whitest of all Silver Pheasants. The upper body has a chalk-white colour with three to four narrow black lines running across in a wavy pattern. The under-parts, the chin, the throat and the long crest are a magnificent, glossy, deep bluish-black. The rectrices are very striking because of their great length. The central pair is pure white and is decorated on the outer web by a few narrow broken black lines.

The cock Silver Pheasant assumes the adult plumage only in the second year. Because the immature dress is considerably different from the full dress, a short description will be given.

The young cock does not have the pure white upper-body, but is finely vermiculated white, reddish yellow and black. The tail has not yet attained its full length and resembles the tail of the hen, though it is already somewhat longer and more coarsely vermiculated. The crest and the under-parts are a dull black. V-shaped white lines are running across the under-parts.

HEN. Though the hens show a great deal of individual variation, it may safely be said that they are mainly olive-brown with more or less inconspicuous black vermiculation. Chin and throat are spotted with grey and the crest is tipped with black.

Sizes:

COCK: length 1,200–1,250 mm (48–50 in.); wing 265–297 mm (10·6–11·8 in.); tail 600–750 mm (24–30 in.); culmen 27–35 mm (1·1–1·4 in.); tarsus 95–105 mm (3·8–4·2 in.).
HEN: length 700–710 mm (28–28·4 in.); wing 240–260 (9·6–10·4 in.); tail 240–320 mm (9·6–12·8 in.); culmen 26–28 mm (1–1·1 in.); tarsus 85–90 mm (3·4–3·6 in.).

The familiar Silver Pheasant inhabits the highland forests, the bamboos and the brush in the south of China. Especially in the province of Kwangsi, it is very common up to heights of 5,000–6,000 feet. In a wild state the male is polygamous. The wild Silver Pheasant is not shy.

This pheasant was mentioned as long as five thousand years ago in Chinese poetry, and its image has been a religious symbol.

It was introduced into Europe in the far-distant past, and was first reared there about the year 1700.

The attractive size of these pheasants, their splendid plumage and beautiful shape, allied to tameness, absolute hardiness and inclination to breed in captivity, make them very desirable aviary birds. They have long been most popular and have been found in great numbers.

One cock can be associated with several hens. The only drawback is the pugnacity of the cocks towards other birds. They may even attack their keeper. Among themselves, however, they are less quarrelsome than many other pheasants. If they are given enough space, several cocks and hens can safely be kept together.

Unlike the genus Lophura, just described, the Silver Pheasant is mature only in its second year, though occasionally a one-year-old hen may lay some eggs. The Silver Pheasant hen is a trusty brooder; she takes twenty-five days to incubate her six to eight rosy, buff-coloured eggs.

THE IMPERIAL PHEASANT (*Lophura imperialis*)

Imperial Pheasants are extremely rare, and only a very limited number are in captivity. A short description (based on the detailed description by Delacour) will, therefore, be necessary to give readers an impression of the appearance of this bird.

FIG. 8. Imperial Pheasant.
(*Lophura imperialis*)

COCK. The colour of the entire body is dark blue. The body-feathers are black with a broad blue fringe. Those of the lower back and rump, the wing and tail-coverts are deep black and have bright metallic-blue borders. The blue-black crest is rather short and pointed. The long, slightly curved central rectrices are pointed and have brown spots, as well as the back and wings. The skin of the face or wattle is scarlet, the legs are crimson. The iris is a reddish orange and the bill is a pale yellowish green, the base blackish.

The cocks assume the adult plumage at the age of eighteen months.

The young males are dark brown with some feathers on the back showing a black border. Tail, crest and spurs are still short.

HEN. There is no real crest. The feathers of the crown are somewhat elongated and are often raised up. The head is light greyish brown, the cheeks, chin and throat are a bit paler.

The upper parts are chestnut with inconspicuous black spots, the under-parts are light greyish brown and sometimes slightly mottled.

The sizes of both sexes are:

COCK: length 750 mm (30 in.); wing 252 mm (10 in.); tail 300 mm (12 in.); culmen 30 mm (1·2 in.); tarsus 87 mm (3·5 in.).

HEN: length 600 mm (24 in.); wing 214 mm (8·5 in.); tail 190 mm (7·6 in.); culmen 28 mm (1·1 in.); tarsus 67 mm (2·7 in.).

The Imperial Pheasant is found mainly in the mountains of central Annam and was first discovered as recently as 1923. Up till now only one pair has been captured. This is because the wild Imperial Pheasant is extremely rare and frequents almost impenetrable mountains and dense forests. Consequently little is known about its life-habits, though Delacour has described his experiences with the species. These experiences may be of relatively little interest to English readers, as at present the only specimens of the Imperial Pheasant available are in France and America, but the breeding data contains much of general use in the rearing of pheasants. The following is a summary of Delacour's findings:

"The original pair of Imperials were first housed in a 20 by 14 feet pen with a large shelter. After some time the hen started laying, producing a clutch of seven eggs, one only being fertile. We then gave them the run of a larger, heavily planted, compartment. The hen then laid three more eggs, all fertile, two of which hatched without difficulty and were reared, but one was killed by a polecat and only one cock reached maturity.

"They were brooded by a Bantam hen and kept in a coop with a run moved to fresh lawn twice a day, and later on let out in the field for a few hours. They were fed on custard and insectile mixture."

In 1926 one cock and two females were reared, and gradually more Imperial Pheasants were produced.

It is solely due to the able breeding of Delacour that a number of pairs are at present available. It is unfortunate that they have all been exported to America, but they will no doubt continue to multiply so that in time this magnificent variety of pheasant will be found in this country as well. The fact that Imperial Pheasants proved perfectly hardy in the European climate is a great advantage.

Fertile hybrids were produced by mating the Imperial Pheasant cock with Edwards's, Swinhoe's, Silver and Black-breasted Pheasants. The Imperial is undoubtedly a close relative of Edwards's and Swinhoe's Pheasants; on the other hand, it resembles in its shape and crest the Black-breasted Kalij, so it might be considered to form a link between the different species.

THE EDWARDS'S PHEASANT (*Lophura edwardsi*)

A short description will give the reader an adequate idea of this small and beautiful pheasant.

COCK. The tail is rather short, compared to the other dimensions of the body. It is quite straight except for the two central rectrices, which are somewhat rounded and blunt-edged in shape. The crest is generally white, with a few black-spotted feathers, and it is quite short. The overall colour of the body is dark-blue with broad shiny blue fringes to the feathers. On the lower back, the tail-coverts and the rump, these fringes are preceded by a deep-black border with a green fringe on the wing-coverts. The bill is greenish white, the tarsi are crimson, the iris is reddish brown. The face-wattles are scarlet and have two large, scarlet lobes.

HEN. As most other pheasant hens, the hen Edwards's is mainly brown, though in this case the brown tends to chestnut. This colour is richest on the mantle, and it is dullest—almost brown-grey—on

FIG. 9. Edwards's Pheasant.
(*Lophura edwardsi*)

the head and the neck. There is hardly any crest. The entire plumage is covered by almost invisible black vermiculations. The six central rectrices are dark brown, the others black. The same division of colour can be seen on the primaries. The bill is a horny brown, the legs are scarlet and the iris is brown.

The sizes of the sexes may be compared thus:

COCK: length 580–650 mm (23·2–26 in.); wing 220–240 mm (8·8–9·6 in.); tail 240–260 mm (9·6–10·4 in.); culmen 30 mm (1·2 in.); tarsus 75 mm (3 in.).
HEN: wing 210–220 mm (8·4–8·8 in.); tail 200–220 mm (8–8·8 in.).

The Edwards's Pheasant is found in the damp mountain forests of Central Annam, but, unlike the other pheasants, never higher than 3,000 feet. As it is extremely wary, it is rarely observed in the wild state, so little is known of its habits. Hence for a long time it was considered to be among the most rare species, and only in 1923 did Delacour succeed in catching a few Edwards's. He took them to Europe. Though this bird lives in a tropical climate, it has proved completely hardy. In 1925 the first Edwards's Pheasants were reared. It was found that incubation lasted only twenty-one days, a short period in comparison with the eggs of their nearest relatives, the Imperial and the Swinhoe, which take twenty-four and twenty-five days respectively to hatch.

Fertile hybrids were produced with a cock Edwards and a hen Swinhoe.

After some years of successful breeding of the Edwards's Pheasant, with the introduction of a few wild-caught birds now and again for a change of blood, the continuity in captivity of this originally extremely rare specimen was assured. Today the species is numerous in Europe and America.

These pheasants are also satisfactory in the aviary. They are beautiful, rather small, perfectly hardy and prolific. An added advantage is the fact that they assume the adult plumage the first autumn after birth, though, as a rule, they breed only when two years old.

THE SWINHOE'S PHEASANT (*Lophura swinhoei*), Plate III

This splendid pheasant certainly merits an introduction.

COCK. The colour plate opposite page 48 shows this magnificent bird in all its splendour, so there is no need for a detailed description. The chief characteristic of this pheasant is its crest, which is mainly white but has a few black streaks, similar to the crest of the Edwards's Pheasant. With the exception of an irregularly shaped white patch on the back, its head, neck, back and under-parts are a glossy blue. One recognizes the same pattern of cross-bars that is found with the Edwards's Pheasant on the feathers of the rump, the lower back and the tail-coverts. The Swinhoe's pattern is different from the Edwards's as the black subterminal bar is narrower and the blue border is wider.

The wing-coverts are black with green borders; the scapulars are red, and together with the white patch on the back they are the typical characteristics of the Swinhoe's.

The tail is dark blue apart from the two greatly elongated central rectrices, which are entirely white and pointed. The bill is horny yellow, the legs crimson, the iris reddish brown and the face-wattles are scarlet.

HEN. The face and the throat are dull-white to pale-grey. The body feathers are mainly chestnut vermiculated black. On the breast the black vermiculation is in a V-shaped pattern, fading into the abdomen.

The bill is a horny yellow, the legs are crimson, the iris is brown.

Young cocks have an immature plumage in their first year; not until the second do they have their full dress and reach maturity.

Swinhoe's Pheasants inhabit the forests covering the mountains of Formosa, and little is known about their life-habits in the wild state.

The bird was discovered by the English naturalist Swinhoe in 1862 and the first live pair was imported into Europe in 1866. These birds were so prolific that in ten years their price fell to £10 a couple; this is even more impressive when it is realized that the first pair cost £250.

With price no longer an objection, Swinhoe's Pheasant became a popular aviary bird, as it combines a beautiful appearance with perfect hardiness and a capacity to produce excellent breeding results, even in a small pen. The female Swinhoe starts laying early in the year. Incubation lasts twenty-five days. The chicks grow fast and are easy to raise. But adequate protection must be provided against contagion, as they are susceptible to infections.

It has been mentioned that the Swinhoe's Pheasant assumes adult dress in the second year only, but as a result of the long years of breeding in captivity—it is more than seventy years ago that freshly caught birds were imported from Formosa for a change of blood—Swinhoes of both sexes which will breed the first year after birth are sometimes found.

The Firebacks

THE Firebacks, which owe their name to a bright red patch on the lower back and rump, are divided into two groups, dependent on whether they are crested or crestless.

The Crestless Firebacks are very rare. Practically nothing is known about their existence and few specimens have ever been introduced into Europe. One or two pairs of pheasants belonging to this group have been in a pheasantry for short periods, but no satisfactory results were ever obtained with these birds. Consequently, the Crestless Firebacks, beautiful though they may be, are of no importance to the pheasant-breeder in this part of the world. To mention their names only, the species known at present are:

Salvadori's Pheasant—*Lophura inornata inornata*;
Malay Crestless Fireback—*Lophura erythrophthalma erythrophthalma*;
Bornean Crestless Fireback—*Lophura erythrophthalma pyronota*;
Atjeh Pheasant—*Lophura inornata hoogerwerfi*.

The Crested Firebacks are more common in pheasantries and more is also known about their life-habits in the wild state. After the last war a few were shipped to Europe, and although at present birds of this species are rare, in time they should be more generally found. They live wild in Malaya, Sumatra, Banka and Borneo.

Both sexes have a short, thick crest of many stiff feathers. In the females the crest is much less developed than in the males. This pheasant is still too rare to make it necessary to go into great detail about it or to include a colour description.

At present four species of the Crested Fireback are known, viz.:

Lesser Bornean Crested Fireback (*Lophura ignata ignata*)

In the aviary these pheasants do well as a general rule, but they should be sheltered against cold. They cannot stand frost, and in western Europe, with frequent wet, cold winters, they soon begin to look unwell.

Because of the cold winter, the hens lay late, so that the chicks are

Fig. 10. Lesser Bornean Crested Fireback.
(*Lophura ignata ignata*)

not sufficiently grown up when the weather turns cold, and hence are not in a fit condition to withstand it. Under very favourable conditions the hen will lay about twenty eggs and the chicks will grow rapidly. They are, however, most susceptible to many germs.

Greater Bornean Crested Fireback (*Lophura ignata nobilis*)

Like the Lesser Bornean Crested Fireback, being only slightly different in size.

Plate V

Reeves's Pheasant
Syrmaticus reevesi

Silver Pheasant
Lophura nycthemera nycthemera

Elliot's Pheasant
Syrmaticus ellioti

Golden Pheasant
Chrysolophus pictus

Lady Amherst Pheasant
Chrysolophus amherstiae

Malay Crested Fireback (*Lophura ignata rufa*), Plate III

Often called Vieillot's Crested Fireback, it is heavier in build than the two preceding Firebacks, and perhaps rather less spectacular, but this does not prevent it from being a magnificent bird. Generally speaking, the Vieillot's Pheasant is easier and better to raise than the Bornean Firebacks. It is essential, however, to provide a closed night shelter in winter. If the cold is severe or continued the run should be shut off with glass; for successful rearing heat is absolutely necessary. Sudden changes in temperature and also damp may prove fatal.

Young Firebacks start looking adult in their first year, but they breed only the second year. When well tended and efficiently housed they are easily acclimatized, and even become strong birds. It should be remembered that their feet are susceptible to frost; temperatures of more than 7° C. below zero are fatal.

At present practically no Vieillot's Crested Firebacks are to be found in Europe.

The Siamese Fireback (*Lophura diardi*), Plate III

Siamese Firebacks are much easier to keep as aviary birds than any other of the Firebacks hitherto discussed. They are far less susceptible to cold and frost, and when properly acclimatized they are hardy even in our inclement climate. Damp and snow are bad for them, however, and it is therefore better to lock them up if the weather is severe. As they are a hardy species, it is not necessary to heat their night shelter; but it is advisable to cover the floor of their night house with a thick layer of straw so as to prevent deformation of the feet caused by too low temperatures. They are monogamous, but care should be taken during the mating-season, as the male is apt to pursue the female too much, and in view of its long and sharp spurs this may prove unpleasant.

The chicks are quite easy to rear, but they must be sheltered from the sun. They are also dainty eaters (this applies to the chicks of all other Firebacks), making it strictly necessary to provide a choice of food; live insects are a vital ingredient of their diet.

Chicks of Firebacks are extremely susceptible to various germs (such as paratyphoid), and should be innoculated if possible.

The Game Pheasants
(Phasianus)

A VERY numerous and widely distributed genus of pheasants is formed by the "*True*" or "*Game Pheasants*". The latter name, which is also used as the title of this chapter, arises from the fact that they make excellent game-birds. This way of describing them has become so common, and is so generally accepted, that the correct title of "True" pheasants has become completely meaningless to many people.

The wide scope of the genus Phasianus and the extent of its geographical distribution are fully realized when it is remembered that the genus consists of thirty-one species and that it is found in the natural wild state as far west as the Caucasus and as far east as Manchuria and Japan. In addition, various races of wild Game Pheasants have been successfully introduced into large parts of Europe and America, where they have become perfectly acclimatized and have multiplied greatly.

Their history goes back a long way and they are the oldest and best-known of all pheasants. In a wild state the various races are found in neighbouring, and sometimes even partially overlapping, areas. They cross-breed easily and hybrids are very fertile, hence many intermediate varieties are found in intermediate areas of two races.

As the purpose of this little book is to stimulate the keeping and raising of ornamental pheasants, the game pheasants should not be omitted, but, on the other hand, there is no need to make a thorough study of all the thirty-one different species. Attention, therefore, will be focused on the few that are important, without becoming involved in extensive and elaborate colour descriptions. Some general information will be given about the genus Game Pheasants and short individual descriptions will follow.

When comparing individually the different species one is struck by the close resemblance of all hens; the variations appear mainly in the cocks. The males are divided into two distinctly different groups. There are the pheasants living on the Asiatic continent and the island of Formosa; they all have a coppery-red or yellow mantle, sides of breast and flanks more or less barred with metallic black, purple or dark green. Then there are the pheasants inhabiting Japan; they are entirely green on the mantle and the under-parts. Based on this striking difference, the genus Phasianus has been subdivided and the species are known respectively as *Phasianus colchicus* and *Phasianus versicolor*.

When it is remembered that the pheasants belonging to the *Phasianus colchicus* are geographically distributed over an enormous area from the Caucasus to the east of China it is not difficult to imagine that many geographical variations of form will appear. An interesting peculiarity is that these geographical variations are all purely breeding birds, and consequently should be considered definite species. This also explains the great number of different species.

The Game Pheasants, unlike all other related genera of pheasants, live in open, flat country. In the wild state the nest usually consists of a hollow in the grass, sheltered by a low bush or brush wood. As a rule, laying begins at the end of March or the beginning of April, and the number of eggs in a clutch varies from fifteen and seventeen. Incubation lasts twenty-three days. The hen is a good and faithful brooder, but she should be disturbed as little as possible, especially in the first few days. The rearing of the chicks presents no problems, though they require large amounts of insects during their first days. Game Pheasants acquire their complete adult dress in the first year.

The males are polygamous and pay not the slightest attention to the incubating female or to the chicks.

Game Pheasants make good aviary birds, though they are always shy. Because of their habit of flying up abruptly when their quarters are approached, pheasants with damaged heads are often found in the aviary. If these captive birds are kept purely as ornaments they must either be tamed or be pinioned.

The many species of the genus Phasianus are subdivided into a few groups according to common characteristics. Here are short descriptions of them.

The Black-necked Pheasants

The four sub-species of this group all have light-brown to buff wing-coverts and are but slightly different from each other. There is no trace of a white ring round the neck.

Fig. 11. Southern Caucasian Pheasant.
(*Phasianus colchicus colchicus*)

The Black-necked Pheasants have been much persecuted, and it is an open question whether they will survive in a wild state. Originally the Greeks and Romans introduced them into Europe, where they were generally established before the tenth century.

They are still found there, though they have become much crossed and mixed with other imported species.

The Black-necked Pheasants are divided into:

Southern Caucasian Pheasant (*Phasianus colchicus colchicus*), Plate I
Northern Caucasian Pheasant (*Phasianus colchicus septentrionalis*)
Talisch Caucasian Pheasant (*Phasianus colchicus talischensis*)
Persian Pheasant (*Phasianus colchicus persicus*).

THE WHITE-WINGED PHEASANTS

These birds, originally inhabiting Turkestan, northern Afghanistan and western China, closely resemble the Black-necks, but are

FIG. 12. Prince of Wales's Pheasant.
(*Phasianus colchicus principalis*)

easily distinguished by their pure-white wing-coverts. Some of the species of the White-winged Pheasants possess a white neck-ring; in others it is just indicated.

The White-winged Pheasants are still fairly abundant in their home country, partly because several of the six sub-species breed in somewhat inaccessible regions.

The White-winged Pheasants are divided into:

Prince of Wales's Pheasant (*Phasianus colchicus principalis*)
Zarudny's Pheasant (*Phasianus colchicus Zarudnyi*)
Khivan Pheasant (*Phasianus colchicus chrysomelas*)
Bianchi's Pheasant (*Phasianus colchicus bianchii*)
Yarkand Pheasant (*Phasianus colchicus shawi*)
Zerafshan Pheasant (*Phasianus colchicus zerafschanicus*).

The Kirghiz Pheasants

The Kirghiz or Mongolian Pheasants are large birds found from the Sea of Aral to the north-western parts of Chinese Turkestan. They generally resemble the birds of the preceding groups, but may be distinguished by their strong green plumage. The neck is circled by a broad white collar, interrupted in front. The true Kirghiz Pheasant (*Mongolicus*) is the only one of this group that was introduced into Europe and America; it is now well established and quite common because it is so strong and prolific.

Three species are known, viz.:

Syra Daria Pheasant (*Phasianus colchicus turcestanicus*)
Aral Pheasant (*Phasianus colchicus bergii*)
Kirghiz or Mongolian Pheasant (*Phasianus colchicus mongolicus*)

The Tarim and Grey-rumped Pheasants

Now follow two groups of Game Pheasants, the Tarim consist of one species, the Grey-rumped of seventeen different sub-species. Most of them are of no interest to us because they were never imported into Europe. We shall therefore discuss only the Chinese Ring-necked Pheasant (*Phasianus colchicus torquatus*), Plate I, and a dark mutation, the Melanistic Mutant Pheasant (*Phasianus colchicus mut. tenebrosus*).

The Tenebrosus are a dark and very beautiful mutation of the species Phasianus. Their actual origin is unknown, but it is generally assumed that this mutation took place in Japan. In any case the Melanistic Mutant was found in England as early as 1880 in the collection of Lord Rothschild.

The bird has dark, entirely metallic, green upper parts, except the wings and abdomen, which are blackish olive. The breast and

FIG. 13. Mongolian Pheasant.
(*Phasianus colchicus mongolicus*)

sides of the body are purplish blue, the neck is also bluish. Since 1933 this dark mutant has been quite common in Europe and America. Today we regularly find entirely deep green and blue cocks and black hens.

The mutation may be considered to have reached its final colour and to be stabilized. These dark pheasants look very handsome; they are also large, well acclimatized and prolific birds.

THE GREEN PHEASANTS

The pheasants described in the preceding chapter all have their origin in the Asiatic continent. The Green Pheasants, however, are found in Japan. As a consequence of their isolated area of diffusion, they have some striking points of difference from the "continental" pheasants, notably the green crown, blue throat, purple neck and green mantle, back and rump. The tail has also a green colour, tending to olive-green, and is patterned with black vermiculation. The under-parts seem to be entirely green as well, but closer inspection shows that some of the feathers are blue and purple.

The Green Pheasants are divided into three races, though numerous local variations occur, which might be called geographic species. Here only the three pure-breeding main races will be mentioned.

Green Pheasants do quite well in their aviaries, but they are usually wild, and consequently more difficult to raise than the other Game Pheasants. Pure specimens are still available in Europe, though many have lost their purity of race by crossing with other members of the Phasianus. The three races belonging to the Green Pheasants are:

Southern Green Pheasant (*Phasianus versicolor versicolor*), Plate I
Pacific Green Pheasant (*Phasianus versicolor tanensis*)
Northern Green Pheasant (*Phasianus versicolor robustipes*).

The Monals
(Lophophorus)

THE heavy build of this species and the short, strong legs with their fearsome spurs make them stand apart among the other breeds of pheasants. The bill, too, is quite different from that of the other varieties. Its peculiar shape—the upper mandible is elongated and curved—makes it an excellent tool for digging.

The tail consists of eighteen feathers and is of medium length. Again the shape is quite unique. The Monal's tail is broad, flat and almost square. It is even shorter than the rounded wings.

The cocks assume the splendid adult plumage only in the second year. First-year males resemble the females, but may be distinguished by the presence of *black spots* on their white throats.

The Monals are true mountain birds; they are found in the eastern part of Afghanistan, the Himalayan Range and the mountains of western China.

Three distinct races of the Monals exist, but their close relationship is so obvious that there is no ground for a generic distinction among them. Hence modern nomenclature unites all three Monals in the genus Lophophorus, and all other names still in use are considered as synonyms. The three known species of Monals are:

Himalayan Monal (*Lophophorus impeyanus*), Plate IV
Chinese Monal (*Lophophorus lhuysi*)
Sclater's Monal (*Lophophorus sclateri*).

The two latter Monals, however, are very rare and extremely expensive, if, indeed, they can be obtained at all. They will not be discussed further in this book, but some information will be given about the Himalayan Monal. Before doing this, however, some general data about the genus Lophophorus may prove helpful.

The colours of the Monal are overwhelmingly beautiful, and the bird may be considered one of the most magnificent representatives of the Pheasant family. The only birds equalling the many-coloured reflections—blue, purple, green, red—of the plumage of the Monals are Humming-birds and Birds of Paradise. Some pheasant-fanciers are of the opinion that the heavy build of the Monals detracts from their beauty, but others consider the size of the bird an additional merit.

Monals are found at high altitudes, from a minimum of 6,000 feet up to the limit of trees at 13,000–15,000 feet. They inhabit open forests of conifers and oaks and move up and down with the seasons. Their long, curved bill is used almost constantly in the quest for food—to uncover grubs, roots and bulbs, which comprise their favourite menu. This pheasant never scratches the ground with its feet.

The Monal is hardy and is not bothered in the slightest by snow and ice.

The display of the cock Monal is aimed at showing his wealth of colour at the fullest. At the same time the bird tries to hid his less attractive aspects from the hen, such as his heavy legs. It is interesting to note that at all times during the display ritual the cock keeps one wing lowered, hiding his legs and feet from the hen. At the beginning of the display the cock approaches the hen in wide circles, with long, careful steps. Important details in the cock's display are the raised crest, which is continually moving, and the outstretched neck. The bill is pressed down against the throat. When the cock starts picking at the ground in front of the hen, making a few rapid bows with spread wings, the excitement is at its highest. This behaviour is strongly reminiscent of the ceremony of the Peacock, especially the position of the wings, which are partly opened and kept at the sides of the body. The feathers of the mantle and the neck are raised to their full length and stand in a circle around the head. The feathers of the wings and the back are hardly raised at all. Particularly during the display the cock Monal with its rich colours is an impressive spectacle.

Lady Impey was the first to keep Monals in captivity. The first breeding results are mentioned in the years 1854 and 1856.

A few decades went by, at first with great successes, but later serious disappointments occurred. In spite of many set-backs, however, a considerable number of Monals were raised, though on account of their adaptation to high altitudes, they proved extremely susceptible to a variety of infections. Soil suitable for Monals must be of a special texture, and accommodation has to meet specific requirements. Rain does not bother these birds, but stagnant damp is bad for them. In countries with a damp climate (such as the north of France, England and the Low Countries) and a heavy soil, special precautions must be taken to ensure the survival of the Monals and to raise their chicks.

Professor Ghigi, one of the greatest experts in the rearing of pheasants, writes in his *Monografia*:

> "Monals are strong and live long, but they are also very subject to germs, more so than most Pheasants. I often obtained perfect specimens, but I never could keep them more than four or five years. They always died suddenly of infectious diseases."

Monals start to lay their eggs at the beginning of April, or, if the spring weather is very favourable, in the latter half of March. A normal clutch numbers four to eight eggs, of a creamy-white colour and richly sprinkled with reddish-brown spots. A good pair will continue reproduction for thirty years—as practical experience has taught. The food of Monals is much like that of the other pheasants, and may consist of a mixture of Indian corn, wheat and buckwheat and much greenstuff. In winter the green food may be replaced by the roots of wild endives and dandelions or carrots cut up into pieces.

From 15 March to 15 June the Monals should be given stimulating food. Mr. H. Flocard, the most successful breeder of Himalayan Monals, writes on this subject:

> "It is best to feed twice daily bread crumbs and hard-boiled eggs, a single egg to a pair of birds, or two if there be an extra hen. A larger amount of this egg food will prove injurious and soon the birds will refuse even this small quantity.
>
> "A conifer should be placed in the aviary, a fir, arbor-vitae or yew, at the base of which an excavation should be made containing an artificial egg. Here the hens will lay.
>
> "The eggs should not be allowed to remain long in the nest for fear of accident. They should be placed in an uncovered box, on a bed of wheat or other grain, in a place damp rather than dry. The hen will lay about six eggs, at intervals of two, three, four or even five days. When the set is completed it will be noticed that she will spend the night on the nest. Both the artificial egg and any of the Impeyan which may be in the nest should now be removed and the nest hollow completely covered for a period of about twelve days. Then by uncovering the nest and replacing the artificial egg a second laying may be induced, and in a similar fashion even a third set of eggs.
>
> "By this procedure twelve to fifteen eggs can be obtained each year from each hen. After 15 June no more eggs may be expected.
>
> "The eggs should be placed under a broody hen, the time of incubation being twenty-seven days, and not thirty as many authors state. The hen should be shut up with her chicks for several days and then liberated in the aviary. Hard-boiled eggs,

bread crumbs and ants' eggs should be provided for the young birds. It is well to supply quantities of living larvae of ants and other insects. The drinking water should be boiled until the birds are three months old."

As we have mentioned, the soil should conform to certain standards. Damp fenland or rich clay is far from ideal, and even harmful. Monals are best kept on a coarse, sandy soil. They need a roomy pen. An area of 150 square feet is the minimum in which to obtain satisfying breeding results. It is also essential to plant dense cover in the aviary. The birds like to have plenty of cover and space, and this induces them to breed more easily. In our wet climate Monals should not be allowed to sleep out in the open. A shelter of approximately 6 feet by 6 feet provided with perches should be installed. It is often found that these birds prefer a broad board to perches for roosting. The chicks, too, which have to stay inside in wet weather, should be accustomed as early as possible to roosting on a perch under a shelter, though they should not sleep outside the coop until they are at least seven weeks old. By the time they are some six weeks old one may start feeding them small grains such as millet, canary seed, a little hemp and buckwheat, but the chicken-mash should be continued at the same time.

To sum up, it may be said that Monals are fairly easy to breed if suitable accommodation is provided and their special requirements are taken into account. The food needed by the chicks is like that of all young pheasants, but with an abundance of live insects and green food.

Finally, it is possible to keep the cock Monal with more than one hen if the aviary is very spacious—at first great caution is essential—but several cocks should never be placed together, unless no hens are around.

The Himalayan Monal or Impeyan (*Lophophorus impeyanus*), Plate IV

After the rather extensive introduction to the habits and keeping of the genus Monal, a few lines about the care of the different species will suffice. It has been said already at the beginning of this chapter that two of the three species, the Chinese and the Sclater's Monal, will not be discussed, as they are extremely rare.

More details will be given about the third variety, but no further colour description, as this species is also far from common in the average pheasantry.

The Himalayan Monal is found in a wild state in eastern Afghanistan, the Himalayas and southern Tibet. It lives at altitudes between 8,000 and 15,000 feet. It is still quite common in the natural state, unlike many other pheasants that were greatly reduced because of increasing cultivation and civilization of their area of distribution.

Everything said in the preceding pages about the genus Lophophorus applies also to the Himalayan Monal, the only one of the three species which has been imported regularly into Europe and reared successfully outside its natural habitat. Today, both in Europe and America, a reasonable number of Himalayan Monals are raised, so there need be no fear that this magnificent pheasant will disappear from our aviaries.

The Cheer Pheasant

(Catreus Wallichi)

Plate VI

WHEN comparing the Cheer Pheasant with the other varieties it is quite evident that this genus is unique among the pheasants. Though there are superficial similarities between the Cheers and the Long-tailed and Common Pheasants in plumage and in build, the bill, the crest, the legs, the voice, the posture and the behaviour all are entirely different. Another remarkable aspect of these birds is their inborn tameness, as opposed to the savage and shy character of all other pheasants. There are even more points of divergence. The colour and shape of the eggs and the chicks stand apart among the other varieties. The cock and the hen resemble each other very much. Considering all this, there is quite evidently no relation whatsoever between the Cheer Pheasants, on the one hand, the Long-tailed and Common Pheasants, on the other.

The Cheers have some characteristics in common with the Eared Pheasants and, stretching the imagination, with the Gallopheasants. But these are of little account, and it is quite certain the Cheers deserve their own specific niche in the Family of Pheasants.

During their first year the Cheer Pheasants assume the adult plumage, though lacking in lustre when compared to older birds. As with other pheasants that have their full dress in the first year, the Cheers are able to breed in the season following their birth.

The area of distribution covers the western and central Himalayas.

This variety is not very common in pheasantries, and it is therefore unnecessary to include a colour description of it. Though the hen resembles the cock, the former is smaller, as is shown by their respective proportions:

Plate VI

Indian Peafowl
Pavo cristatus

Green Burmese Peafowl
Pavo m. spicifer

Cheer Pheasant
Catreus Wallichi

Blue-Eared Crossoptilon
Crossoptilon auritum

COCK: length 950–1,000 mm (38–40 in.); wing 235–270 mm (9·4–10·8 in.); tail 450–480 mm (18–19·2 in.); culmen 25–29 mm (1–1·2 in.); tarsus 74–78 mm (3–3·1 in.).

HEN: wing 225–245 mm (9–9·8 in.); tail 320–470 mm (12·8–18·8 in.); culmen 24–27 mm (1–1·1 in.); tarsus 60–63 mm (2·4–2·5 in.).

A clutch of eggs usually contains nine to fourteen pale yellowish-grey, reddish-brown, spotted eggs which are incubated in twenty-six days.

The Cheer Pheasant has a curiously restricted distribution and is found only at altitudes between 4,000 and 10,000 feet.

In winter the birds migrate downwards from the highest regions. Nesting time is from April to June. The Cheer Pheasant is monogamous and the cock is said to assist in the care of the young.

In a wild state the diet of the Cheer Pheasants consists of the same ingredients as the food of the other varieties, i.e. roots, bulbs, grubs, seeds and berries. Grasses and leaves are not a regular part of their menu. The way in which they obtain their food is quite remarkable: they dig for it almost exclusively with their curved beaks, a habit they have in common with the Monals and the Eared Pheasants. The Cheers seem to have little need for drinking water, consequently they may be found in very dry regions.

The Cheer Pheasant prefers to live on the ground, and even spends his nights not in a tree but sitting on the ground.

There is hardly any display. The entire courtship is very simple, reminding one of the behaviour of the Eared Pheasants; this is one of the few characteristics the Cheers have in common with that species.

The first mention of Cheers being introduced into Europe dates from 1857, and young were first raised in 1858. But it soon became apparent that the change-over to captivity was not without difficulties, as the Cheer Pheasant, too, is susceptible to germs and dampness. However, a considerable number of the species were bred, and from then on in the years 1860–80 large shipments were made from India to Europe. During the First World War the Cheer Pheasants disappeared completely, and it was not until 1933 that they were again found in Europe. About that time pairs were also imported into America, and today it may be said that Cheer Pheasants are well established as aviary birds, though they are much more numerous in America than in Europe.

In the aviary Cheer Pheasants are easy to keep, provided they are on dry ground and given adequate shelter against rain. They can stand any amount of cold, but they are severely tried by damp, and by diphtheric and other germs.

Their tame and peaceful nature makes it quite possible to keep them in fairly small pens; but their digging activities should be taken into account. They may be given the same food as the other pheasants, with a greater proportion of proteins. Their lack of interest in greenstuffs has already been pointed out. In Delacour's experience they show an inclination to be carnivorous. He found this out because a pair in his aviaries killed and ate all the smaller birds that they could catch.

It has already been mentioned that Cheer Pheasants are monogamous and, as a rule, the cock and the hen live well together. But it may happen during the breeding season that the hen has to be protected from the cock. When new pairs are formed we may see the reverse happen, and the hen may attack and bully the male. It is, therefore, best to keep the two birds of the prospective pair in adjoining but separate compartments for some time, so that they may get to know each other.

Their chicks grow quickly and well, and require the same food as the young of the other pheasants.

Hybrids were obtained of the Cheer Pheasant with the Monal and the Reeves's Pheasant.

The Eared Pheasants

(Crossoptilon)

It is not intended to discuss the Eared Pheasants in detail; only those specimens which are most likely to be found in aviaries will be described. An exception will be made in the case of one of the species of the White Eared Pheasants. Though this is a very rare bird, some space will be devoted to it. In this way all three colour types will be included, for we know three colour types of the Eared Pheasants—viz. the White, the Brown and the Blue Eared Pheasant.

The Eared Pheasants, when compared to the other varieties, appear to have very little in common with most other pheasants, as was the case with the Cheer Pheasants. There is a slight resemblance to the Gallopheasants with regard to the shape of the face-wattles and some peculiarities in the behaviour of the bird. The shape of the bill, however, is similar to the bill of the Cheer Pheasant.

The build of this variety is obviously adapted to life at great heights; the Eared Pheasants are found on the mountain slopes and plateaux of south-eastern Tibet and West and North China, as high as the snow line. They are accustomed to life in barren country, though they need some amount of grass, trees and shrubs.

Hybrids have been obtained with the Monals as well as with the Silvers and Nepal Kalijs. Though it has been stressed that the Eared Pheasant is very much an independent species of the Phasianidae, the successful breeding of hybrids of which the cocks proved fertile shows a definite relationship to the races mentioned above.

The face-wattles with their erectile papillae are reminiscent of the Gallopheasants and the Common Pheasants.

The tail has elongated, almost "hairy" ends and has a compressed shape. The resemblance of male and female Cheer Pheasants was stressed; the resemblance of Eared Pheasants of both sexes is so great that it is hard, for the amateur, to tell them apart. There is no evidence of any difference in the plumage of either sex. The cock has a short spur on the tarsus and is a little heavier. His legs are a bit sturdier and the face-wattles are more rounded and a brighter red, but that is all. Sometimes the cock does not have the spurs, or, as often happens with birds bred in captivity, the hen has spurs as well, and it really becomes almost impossible to tell the sexes apart, in particular with the Brown Eared Pheasant. The lack of difference is all the more surprising when one considers the highly ornamental plumage which implies a strong divergence between cock and hen in the other species of pheasants.

Eared Pheasants are poor fliers and when disturbed usually run uphill and then embark on a heavy but very fast downward flight to the bottom of the valley. By nature they are unsuspecting, but in northern China, where for many years the Brown and Blue species have been much persecuted because of their plumage, they have become wary. In Tibet, on the other hand, where they have never been ravaged, the Eared Pheasants have remained tame. This is due mainly to the protection given, for religious reasons, by the Buddhist population to this species. It is unlikely that these conditions will remain under a Chinese Communist regime.

Eared Pheasants feed on all kinds of seeds and leaves, but they mainly dig with their powerful beaks for bulbs, roots and insects, of which they are fond, seldom, if ever, scratching.

Even at the height of winter the Eared Pheasants remain in practically the same surroundings as in summer. They are sedentary birds which live together in large flocks and consequently have a strongly developed sociability. They do break up into pairs in spring but even then few fights occur. This variety is undoubtedly the most tolerant of all. The hens nestle on the ground on a protected spot, and the eggs are usually laid in May. The Eared Pheasant does well in the aviary or free in a park. A hen may lay a total of thirty eggs in several clutches. Though their original breeding-grounds are at high altitudes where the air is much thinner, Eared Pheasants are easy to acclimatize. They are completely hardy and can stand any degree of cold, but long periods of damp are harmful.

Eared Pheasants live in pairs. If they are to be kept in perfect condition they need a good deal of space. They do better in a park than in an aviary. In large spaces with plenty of grass and adequate bushes they keep their plumage perfect and live a long time. Their great tameness and disinclination to fly make it possible to house them in an open run without a wire-netting roof. But in small pens they quickly start pecking at each other's feathers—particularly the tail-feathers. This bad habit is caused by a lack of cellulose in their diet and by boredom through inactivity. An abundant supply of greens and roots may sometimes obviate this trouble, but the best solution is to keep the hen and the cock in separate pens.

It is understandable that cock and hen should be brought together a number of times during the mating season until it is quite certain that there has been impregnation. But there are more advantages to a separate accommodation for either sex. The cock will not be able to pick the eggs—a bad habit that is quite regularly found with pheasants—and it will be possible to mate several hens, if available, to one cock. The monogamous habits of some varieties will not interfere, as has been shown in countless instances when breeders obtained excellent breeding results with one cock and several hens.

Feeding and general care involve no special requirements.

The chicks grow well and are easy to rear. Only in very damp surroundings the rearing of the chicks may be difficult, as they are highly susceptible to diphtheria. Should there be no alternative but to breed the chicks in a damp place, it will help to add one part per thousand of corrosive sublimate to their drinking-water. When the chicks are three to four months old they begin to look like their parents and it will be possible, with luck, to determine their sex. Eared Pheasants only begin breeding in their second year, though in captivity as a result of inbreeding the birds may prove fertile in their first year.

If enough space is available it is advisable to let the chicks run free from the age of two to three weeks. Their choice of food will become far more varied, apart from a considerable increase in the amount of food they consume.

The Eared Pheasant chicks often have the bad habit of picking at each other's toes. This should be carefully watched so as to avoid a great deal of unpleasantness, or even permanent disfigurements.

But, generally speaking, Eared Pheasants are easy to keep and rear, and they are very easily domesticated.

The White Eared Pheasant (*Crossoptilon crossoptilon*)

Here is a short description of this beautiful, but, unfortunately, extremely rare pheasant.

Cock. The overall colour of the body plumage is white with a grey lustre on the wings and the tail-coverts. There are no conspicuous ear tufts as in the case of the Brown and Blue. The top of the skull is covered with velvety-black feathers, the secondaries are black-brown and the primaries are dark brown in colour. The tail is purple with bronze overtones at the base, shading into dark green-blue and purple towards the tip. The tail consists of twenty feathers; the two central rectrices are longest, and curved towards the ends. They have the same loose, almost hairy construction, though the White Eared Pheasant does not have the disintegrated type of plumes like the Egret, as found with the Brown and even more with the Blue Eared Pheasant. The bill is a horny colour, the tarsi and face-wattles are scarlet and the iris is a yellow-orange.

HEN. Resembles the cock, a little smaller and usually somewhat darker and browner in colour.

Size of the White Eared Pheasant, applying to both the cock and the hen:

Length 920 mm (36·8 in.); wing 330 mm (13·2 in.); tail 575 mm (23 in.); culmen 36 mm (1·4 in.); tarsus 100 mm (4 in.).

The White Eared Pheasant is found in large numbers in south-eastern Tibet, where it has been protected by the religious beliefs of the inhabitants. It lives at great altitudes—between 10,000 feet and the snow-line—but rarely leaves the forests.

With a great deal of effort a few White Eared Pheasants have been

FIG. 14. White Eared Pheasant.
(*Crossoptilon crossoptilon*)

sent over to America, where they have established themselves through successful breeding. They breed fairly well, and the period of incubation lasts twenty-four days. At present a small number of White Eared Pheasants are found in American aviaries, but as this species is less robust and prolific than the Blue and the Brown, it cannot yet be considered as safely established in captivity.

THE BROWN EARED PHEASANT (*Crossoptilon mantchuricum*), Plate III

COCK AND HEN. With the exception of the iris, which has a light reddish-brown colour, the parts of the bird that are not covered by its plumage, such as bill, tarsi and face-wattles, are the same colour as the corresponding parts of the White Eared Pheasant. The crown of the head also has the velvety-black feathers. The tail, however, consists of twenty-two feathers and is dull-white, shading into brownish-black with a purple-blue gloss. The tail resembles the plumes of the Egret, being very loose, with the exception of the two outer pairs. Here again, the central rectrices are longest and most disintegrated. They are curved towards the tips, much longer than the other tail-feathers and carried erect. The chin, the throat and the other feathers on the face are creamy-white. The ear-coverts are very long and chalk-white. The neck is black, gradually shading into brown on the mantle. The feathers of the lower back, the rump and the upper tail-coverts are silvery-white; the underparts are brown. The wing-coverts and secondaries have a purple gloss.

The size of the Brown Eared Pheasant is as follows:

Length 1,000 mm (40 in.); wing 306 mm (12·3 in.); tail 544 mm (21·6 in.); bill 32 mm (1·3 in.); tarsus 100 mm (4 in.).

The Brown Eared Pheasant inhabits the mountains of western China, but because of continual persecution it has become rare. Progressive cultivation of the forests has also greatly reduced the numbers of these birds.

The Brown Eared Pheasant was introduced into Europe for the first time in 1864. Then no breeding results were obtained, but after some four years this variety bred so well that they began to be distributed all over Europe. It is interesting to note that the countless Brown Eared Pheasants at present kept and reared both in Europe and in America are all descendants of the one cock and two hens brought to Europe in 1864.

The incubation lasts twenty-six or twenty-seven days. When raising these pheasants one frequently comes across clear eggs, and it often happens that the cock is infertile. Plenty of space may alleviate this trouble, which, of course, is due to extreme inbreeding. The lack of spurs in many of the captive males is another proof of this. Great care should be taken to continue propagating this species adequately in confinement, for it may become before long completely extinct in the wild state.

THE BLUE EARED PHEASANT (*Crossoptilon auritum*), Plate VI

Here is a short description, applying to both male and female, of this type of pheasant:

This variety also has the velvety-black feathers on the top of the head. The iris, the bill, the face-wattles and the tarsi are the same colour as those of the White Eared Pheasant. Again the elongated feathers of the ear-coverts are chalk-white, and the feathers of the chin, throat and sides of the forehead have the same colour. The overall colour of the body plumage is blue-grey. Here the tail consists of twenty-four feathers, some disintegrated like the plumes of the Egret. The four central rectrices are entirely filamentous, and the twenty remaining feathers all are partly disintegrated to some degree.

The four central rectrices are blue-grey, darker towards the tips and with a metallic green gloss that shades into purple. The other tail-feathers are greenish and violet-blue in colour. The extreme five or six pairs are three-quarters white, though the length of the white part may differ considerably among the individual birds. The secondaries are dark-brown with a purple lustre. The primaries are dull brown.

The proportions valid for both sexes are:

Length 960 mm (38·4 in.); wing 290–306 mm (11·6–12·3 in.); tail 490–560 mm (19·6–22·4 in.); bill 35 mm (1·4 in.); tarsus 101 mm (4 in.).

The habits of the Blue Eared Pheasant are similar to those of the two preceding species. This pheasant possesses the most specialized plumage of the genus. The four middle rectrices which form its tail of twenty-four feathers have extremely long and disintegrated barbs, much more so than the other pheasants. Also the ear-tufts are far longer.

For centuries the Blue Eared Pheasant has been much persecuted for the sake of its beautiful feathers, and the numbers of wild birds have been greatly reduced.

The first live Blue Eared Pheasants to be sent from North China to Europe arrived in 1929. Although at first they did not breed very rapidly, gradually this changed for the better, and today they are just as numerous in European and American pheasantries as the Brown Eared Pheasant. They are, in fact, even easier to manage than the latter. Hybrids between the Blue and the Brown Eared Pheasant have been bred a number of times, and the resultant birds seem to be entirely fertile.

The Tragopans

(Tragopan)

IN a popular work like this it is, of course, impossible to discuss all pheasants extensively. Not even all genera have been included, much less the many different species of those genera.

It was intended originally to omit entirely the Tragopans, but here, after all, are a few words about these splendid and interesting birds.

The genus Tragopan stands alone among the other genera of the sub-family of the Pheasants. Five species belong to this genus, viz.:

Western Tragopan (*Tragopan melanocephalus*)
Satyr Tragopan (*Tragopan satyra*), Plate IV
Blyth's Tragopan (*Tragopan blythi blythi*)
Temminck's Tragopan (*Tragopan temmincki*), Plate IV
Cabot's Tragopan (*Tragopan caboti*)

and a sub-species of Blyth's Tragopan, the

Molesworth's Tragopan (*Tragopan blythi molesworthi*).

Of these, only the Satyr Tragopan will be described individually.

It is an overwhelming task to give a colour description of the Tragopan Pheasants. The pattern is so elaborate and the colours are so mixed, that if one does not know the birds it is impossible to form a picture of their appearance from a written description only.

Suffice it to say that the body-plumage is coloured predominantly red with white, grey, brown, black and blue shades, depending on the species. Some peculiar ornaments, typical of the Tragopan, may be mentioned.

The cocks have short crests. Two fleshy horns and large, brilliantly coloured, bib-like throat-wattle or lappet are displayed during the breeding season. The ornaments give the birds a typical look, and the presence of the horns explains the English name of "Horned Pheasants" and the German "Hornfasanen".

The area of distribution of the Tragopans covers Kashmir in the west, across the Himalayas and Burma into central China, where they are found at great heights, varying between 3,000 and 12,000 feet. Their food consists of leaves, berries, seeds and insects.

In contrast to the other pheasants, they are great fliers, and consequently far less earth-bound. They frequently feed in trees, and for nesting they often use old nests made by crows or other birds. But the hen Tragopan sometimes builds in a tree a well-constructed nest of branches and twigs.

It was pointed out that the cock Tragopan has a number of remarkable ornaments at his disposal, none of which are found with other species of the sub-family. Though we do not intend to give an elaborate description of the entire display ritual of the Tragopans, it may be interesting to tell something about these typical ornaments, the "hidden beauty" of the Tragopans. Hidden beauty, because apart from its lavish colours the bird has some features which are only in evidence during the display ceremony. These are two fleshy horns on the head and sort of throat-wattle that is quite large and has an intricate colour-pattern. During the display of the cock these ornaments are visible for just a moment only. This is surprising, since the size of both ornaments is quite important in proportion to the dimensions of the pheasant, the horns reaching a height of 5–7 cm (2·2¾ in.) when upright and the throat-wattle or lappet stretching out to a length of 10–15 cm (4–5⅝ in.) and a width of 5–7 cm (2–2¾ in.).

Tragopans are perfectly hardy, and they require more protection from summer heat than from winter cold. One disadvantage is that these birds do not survive long in small pens. But if provided with large, well-planted enclosures they live to a ripe old age, and satisfactory breeding results are obtained.

The Tragopan Pheasant assumes its full adult plumage in the second year and only starts breeding then. As a rule Tragopans are best kept in pairs, though satisfactory results have been obtained with one cock to two hens.

They can easily be kept together with other birds, even with other pheasants.

Their food consists of the same menu of seeds and grains as required by the other pheasants, but it is *absolutely indispensable* to provide daily large quantities of fruit and greenstuffs. (Cabbage of all kinds is very bad for Tragopans.)

The breeding season usually starts in April. The hen produces a clutch of three to six eggs at daily intervals. Hen Tragopans prefer to lay in an elevated nest, and it is advisable to fix a nest (e.g. a shallow basket) in a tree. They sit well and make excellent mothers. The incubation of the eggs and the care of the chicks may be left to them without the slightest anxiety. To obtain several clutches the eggs should be taken away immediately and incubated in a machine or under a broody hen.

The chicks are soon able to fly, and as early as their second day come to roost with the hen on the perch. Great care must be taken with their food. Bearing in mind the fact that at present there are almost no Tragopans to be found in Europe and that their high price will continue to act as a deterrent to breeders, it is not necessary to go into their care and upkeep any further, except to add that they require many insects, a great deal of greenstuffs and a carefully prepared mash. Actually, there is no need to worry, because Tragopan chicks need practically nothing to eat for the first two days.

THE SATYR TRAGOPAN (*Tragopan satyra*), Plate IV

No colour description of the Tragopans will be given here; all that has been already said applies without exception to the Satyr Tragopan. This species is found in the central and eastern parts of the Himalayas at altitudes of 8,000 and 10,000 feet. In winter it descends to a height of 6,500 feet.

The first specimens came to Europe in 1863, and in the same year young were reared. Satyrs were raised with moderate success and were being distributed over Europe in large numbers when the two world wars brought the profitable breeding to an abrupt end. Today the bird is found only sporadically in Europe, and in America, too, it has become very rare. In view of the current difficulties in importing birds it seems unlikely that more Tragopans will become available for some time.

The Peacock Pheasants
(Polyplectron)

Two species only from the fourteen of the genus Peacock Pheasants will be discussed here, as the others are seldom kept in this part of the world. But first of all it is advisable to give some general information about the genus Polyplectron.

Peacock Pheasants are far less particular about their food than most other pheasants. They may even be considered omnivorous, since they will feed on fruits and seeds as well as insects and raw minced meat. The meat in fact is a very welcome addition to their daily diet and should be provided regularly, though in small quantities. Furthermore, they are given the normal mixture of cereals, fruit, berries and a nutritious mash. The Peacock Pheasant is not interested in greens, so these are usually left out, though the individual preferences of the birds should be taken into account, something that applies to all varieties of aviary birds. Always try first if a type of food is accepted before rejecting it on account of set rules.

Generally speaking, Peacock Pheasants may be considered well suited to life in the aviary. They are not aggressive, and can be kept together with other birds. They are also quite tame. They are presumed to be monogamous, so it is preferable to keep the birds in pairs. One cock may be kept successfully with two or three hens, yet when kept in pairs the breeding results will be far better. This variety has proved to be a good breeding bird, and the chicks are fairly easy to rear. During the first few days they require a good deal of attention, but as they only need a small run, it is not too difficult to watch them carefully. Further on more details about the raising of the poults will be given.

When comparing the Peacock Pheasants with the other varieties described in this book some characteristics are found that are unique in this species. Other features indicate a relation, though distant, to the Ruffed Pheasants, and some points are reminiscent of the Argus Pheasants. The slender build of the most primitive forms of the species resemble the appearance of the Ruffed Pheasants; but their type of plumage has the same consistency as the feathers of the Argus, which is far more important a link. Both the Peacock Pheasants and the Argus have the same lax plumage and simple colour scheme. The hens of both species lay only two eggs in a clutch.

The feathers of the body are mainly brown, vermiculated white or buff. They are adorned with spots of metallic brilliance, like the well-known ocelli on the tail of the peacock. On account of these patches this variety has been given its name. There are no face-wattles, and the crest may be absent as well. Some specimens have a small ruff. The hens are easily distinguished from the cocks, because they are of the slighter build and have no spurs (the tarsi of the cocks usually have two, sometimes one and in some cases three) and also have duller colours. The tail and the tarsi are much shorter too.

Young cocks resemble the hens, but can be told apart because of their heavier posture, longer legs and the beginning of spurs. The cocks assume the adult plumage in the second year.

The display of the cocks is very elaborate. All representatives of the genus Peacock Pheasant have the lateral display. In addition, the cock offers various delicacies to the hen, walking around her in circles, clucking softly all the while.

Those species that possess a broad tail press their body to the ground, spreading mantle, wings and tail into a full circle, with the metallic patches offering a magnificent and colourful sight. This characteristic of the display is found again with the Argus Pheasants.

The Peacock Pheasants live in dense, humid forests. They occur both in the low country and in the mountains, as high as 4,500 feet.

(*Continued on page 129*)

Photo section showing how birds become acclimatized and thrive under properly planned aviary conditions.

FIG. 15. A view of the spacious well-planned aviaries at Leckford Abbas with a Temmincks Tragopan in the foreground.

FIG. 16. Male Himalayan Monal in a planted breeding pen at the Ornamental Pheasant Trust, Great Witchingham, Norfolk.

Fig. 17. Male Temminck's Tragopan at feeding time (Ornamental Pheasant Trust).

Fig. 18. Young Silver Pheasants in the rearing pen (Ornamental Pheasant Trust).

Fig. 19. Male Yellow Golden Pheasant with female (Leckford Abbas).

Fig. 20. Male Amherst (Zoological Society of London).

Fig. 21. Scintillating Copper Pheasant (Leckford Abbas).

Fig. 22. Pair of Bornean Great Argus Pheasants (Ornamental Pheasant Trust).

Fig. 23. Male Mikado (Leckford Abbas).

Fig. 24. Pair of Mikados (Leckford Abbas).

Fig. 25. Edwards's Pheasant (Leckford Abbas).

Fig. 26. Swinhoe's Pheasant (Zoological Society of London).

FIG. 27. Horsfield's Pheasant or Black-breasted Kalji (Zoological Society of London).

FIG. 28. Vieillot's Fire-backed Pheasant (Zoological Society of London).

Fig. 29. Pair of Silver Pheasants (Ornamental Pheasant Trust).

FIG. 30. Monal Hen
(Leckford Abbas).

FIG. 31. Blue Crossoptilon (Ornamental Pheasant Trust).

FIG. 32. Brown Crossoptilon Hen (Leckford Abbas).

FIG. 33. Brown Crossoptilon Cock (Leckford Abbas).

FIG. 34. Side view of Brown Crossoptilon Cock (Leckford Abbas).

Fig. 35. Temmincks Tragopan side display (Leckford Abbas).

Fig. 36. Satyr Tragopan (Zoological Society of London).

Fig. 37. Display by a Great Argus (Zoological Society of London).

Fig. 38. Another view of the Great Argus display (Zoological Society of London).

Fig. 39. Grey Peacock Pheasant (Zoological Society of London).

Fig. 40. Peacock Pheasant lateral display (Leckford Abbas).

FIG. 41. Peacock display (Zoological Society of London).

FIG. 42. Black-shouldered Peacock (Leckford Abbas).

Fig. 43. Male Sonnerats Jungle Fowl (Leckford Abbas).

Fig. 44. Rheinarts Argus at Clères.

They are consequently hardy in the European climate. Still, it is advisable to provide a sheltered run. In the cold and windy climate of Northern Europe no birds can be left without adequate shelter during winter, which they may often need during summer as well.

The laying period starts as a rule in March. If the breeder should wish to rear more chicks than just two (the normal amount of eggs laid in a clutch), the eggs should be taken away as soon as laid. Usually a second laying will follow within the next few weeks.

Delacour gives some useful and interesting hints about the rearing of Peacock Pheasants:

> "The eggs do not keep for long and must be put into incubation within a few days after they have been laid.
>
> "The chicks are at first rather delicate. They have the habit of taking shelter under their mother's tail and of eating only from her bill, refusing to pick up food from the ground. When raised under a bantam hen, care must be taken that it is a gentle, light one which calls and feeds them properly. Otherwise one has to hand feed them at first, offering them insects at the tip of a needle or forceps. We found it sometimes useful to associate one or two Golden chicks with them to show them how to eat.
>
> "But they must not be mixed with too many other young, even of their own species, six being a maximum. They are small eaters and live insects are useful during the first few days or weeks, particularly mealworms."

GERMAIN'S PEACOCK PHEASANT (*Polyplectron germaini*)

COCK. It is impossible to give an adequate impression of the Germain's Peacock Pheasant solely by describing it. Only a general description of the colours will be given; the colour plate will supply the details. The upper part of the body is dark-brown vermiculated yellow. Head, throat and neck are black with narrow greyish-white stripes. The round, violet-blue ocelli are found on the tips of the mantle and wing-feathers. These ocelli are surrounded by a black ring and often have green reflections. There are no ocelli on the back and rump. The primaries are a dull black. Close to the tip of each of the twenty tail-feathers is a brilliant green-black ocellus.

The tail is flat, and rounded at the end.

FIG. 45. Germain's Peacock Pheasant.
(*Polyplectron germaini*)

The underparts of the bird are black with buffy stripes.

Tarsi and bill are blackish brown, iris is brown, bare facial skin red.

HEN. Somewhat smaller than the cock and of a duller colour. The head and the neck are similar, but the rest of the plumage is dark brown streaked and freckled with pale brownish-grey. The ocelli on mantle and wings are triangular.

Measurements:

COCK: length 550 mm (22 in.); wing 180–200 mm (7·2–8 in.); tail 250–320 mm (10–12·8 in.); culmen 200 mm (0·8 in.); tarsus 65 mm (2·6 in.).

HEN: length 480 mm (19·2 in.); wing 160–185 mm (6·4–7·4 in.); tail 220–250 mm (8·8–10 in.); bill 16 mm (0·6 in.); tarsus 55 mm (2·2 in.).

The Germain's Peacock Pheasant is found in the forests of eastern Cochin China and southern Annam. The birds prefer damp places and live both in the lowlands and as high up as 4,000 feet. They breed almost continuously throughout the year. As soon as the young become independent the hen starts laying again.

Germain's Peacock Pheasants were introduced in Europe in 1875 and started breeding the same year. Incubation lasts twenty-two days. Generally speaking, they are slightly more prolific than the Grey Peacock Pheasants, but are not as hardy. In winter they should be provided with a closed shelter. By regular breeding and occasional imports they have become successfully established in America. They are scarce in Europe at present.

THE GREY PEACOCK PHEASANT (*Polyplectron bicalcaratum*), Plate I

COCK. There is no need to describe the colours, as the colour plate gives a very good picture of this splendid bird.

To facilitate making a distinction between the two species of Peacock Pheasants discussed here, the most striking differences will be pointed out.

The ocelli on the mantle and wings are lighter in colour than those of the Germain's and have more of a blue and purplish-red lustre.

The feathers have a white instead of a brown border.

The ocelli on the tail are larger and have a lighter green gloss.

The general colour of the body is lighter and more predominantly grey.

The tail has twenty-four feathers instead of twenty, and is longer and wider.

The iris is white, the facial skin is pale yellow, the tarsi and bill are grey.

HEN. Much smaller than the male. The entire plumage is darker, duller and irregularly marked. The tail, the crest, the ruff and the legs are short.

The Grey Peacock Pheasant lives in the damp forests of Burma and northern Siam, at altitudes of up to 5,000 feet.

As early as 1745 a few specimens were imported into Europe, but these died without having bred. It was not until 1863 that more Grey Peacock Pheasants came to Europe. Soon young were reared, and until the Second World War this pheasant was common in European pheasantries. After the war, however, it became rare in Europe and is seldom found in America.

The Grey is quite hardy, and in our climate little extra care is required, though a shelter is desirable.

These birds are prolific, and the hen is an extremely good mother. Although the young do not assume the full adult plumage until the second year, they sometimes prove fertile after one year.

For further details it is advisable to refer to what has been written about the Germain's Peacock Pheasant.

The Argus Pheasants

THE Argus Pheasants are found only sporadically in private collections, as they are extremely rare and consequently very expensive. They could, therefore, have been omitted from this book, but their magnificent appearance warrants their inclusion. It is unnecessary, however, to go into elaborate details, and a few general remarks will suffice.

The heading "The Argus Pheasants" comprises two genera in one denomination, viz.:

(*a*) the Crested Argus (*Rheinartia*), and
(*b*) the Great Argus (*Argusianus*).

To the genus Rheinartia belong the
Rheinart's Crested Argus (*Rheinartia ocellata ocellata*) and the
Malay Crested Argus (*Rheinartia ocellata nigrescens*).

The genus Argusianus is formed by the
Malay Great Argus (*Argusianus argus argus*), Plate IV;
Bornean Great Argus (*Argusianus argus grayi*); and the
Double-banded Argus (*Argusianus bipunctatus*).

Of all the above-mentioned we shall probably come across only a few specimens of the Great Argus. A few pairs of this splendid bird are to be found in Europe and occasionally a single specimen—and with luck a mated pair—may be discovered in zoological gardens or in the possession of a wealthy private collector. No birds of the genus Rheinartia had been seen alive before 1923, and their existence had been merely guessed at because of old chronicles and a few tail-feathers in the Paris Museum. As a matter of curiosity it may be added that the tail-feathers of Rheinart's Crested Argus are the longest birds' feathers in the world—at least of wild birds. Ancient chronicles relate that these birds were offered to the Emperors of China as ritual presents and that the legendary Chinese Phoenix was created in the image of this Crested Argus.

In 1924 the first Crested Argus Pheasants were brought to Europe by Delacour, who succeeded in breeding from them.

These birds, natives of central Annam, were found to have an

FIG. 46. Rheinart's Crested Argus.
(*Rheinartia ocellata ocellata*)

incubation period of twenty-five days. The chicks grew well on the usual pheasant food and behaved much like young Peacock Pheasants.

There is no need to include a colour description, nor to go deeply into their habits.

No live specimens of the Malay Crested Argus have yet been kept outside their natural state, so they are not described here.

It only remains to be said about the genus Rheinartia that at present no specimens are available, except a few hens in America. Judging from the length of time that Delacour possessed these birds, it is evident that they do well in captivity—his specimens lived for twenty years—and are fairly easily induced to breed. A pair of Crested Argus requires a roomy aviary, heavily planted, with plenty of shade. More than one female cannot be kept in one pen. They must be fed a great deal of soft food, fruit, insects and some raw meat. The birds are not delicate, but they are susceptible to diphtheria; there are, however, enough disinfectants available on the market. Large numbers of these pheasants are still found in a wild state.

THE MALAY GREAT ARGUS (*Argusianus argus argus*), Plate IV

The Great Argus lives on the Malay Peninsula, Sumatra and Borneo, where it is found at low and moderate altitudes. The variety in Borneo is clearly different from those in the two other regions and has been called *Argusianus argus grayi*.

A colour description of the Great Argus is not needed here, as the coloured plate gives a complete picture of its appearance. Only in the display does the Argus cock show his truly marvellous splendour. In ordinary circumstances he seems a rather simple pheasant. But in the display he is quite the reverse.

In a wild state the Great Argus Pheasants are not rare, in contrast to the small number found in captivity. However, they are rarely seen in the wild state, as the birds are excessively wary. They prefer dry, rocky country, at heights from sea-level to about 1,300 metres.

The hens are very good fliers; the cocks move rather awkwardly, due to the long secondaries.

It is known that the Argus Pheasants live separately all through the year. Only during the breeding season does the hen seek out the cock, but probably no longer than the mating period. Great Argus cocks bred in captivity proved dangerous to the chicks. It is advisable to remove the cock from the aviary as soon as the mating is over. The aviaries in which Great Argus are kept should, of course, be very spacious, considering the enormous length of the tail-feathers.

The breeding of Argus Pheasants should not prove too difficult.

The chicks have the disadvantage that they have to be fed "from the bill" during the first few days, a habit that makes it impossible to have them raised by a foster mother, but once they start feeding by themselves the worst is over.

It is evident from the above that the hen Great Argus should brood and rear her own chicks. This means that there will be no second laying, but it is preferable to have two live chicks over any amount of failures. Argus Pheasant chicks grow slowly and have to be given a great deal of care during the first year. They should not be left without artificial heat during the first winter. They fully mature only in their third year, after which the wing and tail feathers continue to grow for another three or four years, before reaching their full length.

Even grown-up Great Argus never get used to our damp winters. An adequate night coop is required.

The Great Argus is a relatively quiet bird, who accepts quite well the company of other pheasants. The hens, however, should never be kept together, nor is it advisable to accommodate Peacock Pheasants in the same aviary.

A great drawback of the Argus Pheasants is the tendency of the cocks to start moulting at the approach of the mating season. Moulting is such a drastic occurrence to these birds that they are sure to be out of condition and quite incapable of impregnating the hen. Careful feeding may somewhat postpone the moult.

The Great Argus, like the Rheinartia, makes a habit of clearing a playground. As a rule this is a circular area with a diameter of 3–5 yards from which every little twig, leaf or pebble is carefully removed. The cock walks up and down on his playground, emitting his call at regular intervals. During the mating season the hens are attracted by the voice of the cock. The extremely complicated and beautiful display ritual takes place in this arena as well.

Sizes are:

COCK: length 1,700–2,000 mm (68–80 in.); wing 800–1,000 mm (32–40 in.); (secondaries 450–500 mm (18–20 in.); tail 1,160–1,430 mm (46·4–57·2 in.); culmen 35–39 mm (1·4–1·6 in.); tarsus 110–120 mm (4·4–4·8 in.).

HEN: length 740–760 mm (29·6–30·4 in.); wing 350–400 mm (14–16 in.); (secondaries 300–350 mm (12–14 in.); tail 310–360 mm (12·4–14·4 in.); culmen 33–36 mm (1·3–1·4 in.); tarsus 85–95 mm (3·4–3·8 in.).

DOUBLE-BANDED ARGUS (*Argusianus bipunctatus*)

Of this bird we possess only a portion of a single primary in the British Museum. There are remarkable differences from the comparable feather of the Great Argus, and there is no doubt that this feather belongs to a hitherto unknown species of the genus Argusianus. All efforts to trace this bird have proved unsuccessful, so it is presumed that the feather belongs to a pheasant that once lived in Java but has now become extinct.

The Peafowls

(Pavo)

THIS book would be incomplete if the Peafowls were not mentioned. Everybody has heard about, or seen, a Peafowl, but not everyone knows that the genus Pavo contains more species than the universally known Indian (Blue) Peafowl, which has its habitat in India. Many keepers and rearers of pheasants will not keep Peafowls as well, largely because of lack of space. Hence only a few general remarks about the genus Pavo will be made here.

Peafowls have been famous for tens of centuries through arts and letters. The Ancient Phoenicians brought them to the Pharaohs of Egypt and Peafowls were known to King Solomon. Today most people have been able to admire them in zoos and parks; with their large size and magnificent plumage they make a fine show.

Peafowls are fairly easy to breed. Only in the third year do they assume their full adult splendour.

There are two species of the genus Pavo which are distributed over Ceylon, India, Indo-China, Malay and Java. They are not found in Sumatra and Borneo.

The Peafowls are polygamous. In the wild state they usually live in small groups of one cock with as many as five hens. The hens nest on the ground and lay four to eight eggs. The incubation takes from twenty-seven to thirty days, usually twenty-eight days. The Peahen is a good mother, and the chicks are easy to rear, though they do not grow rapidly.

Their large size and long feathers give the Peafowls an imposing appearance. Their food is very simple: cereals and greenstuffs are all they need. Moreover, they find a large part of their food for themselves when they are kept at liberty in a park or covert. Peafowls may also be kept in pens, but they should have plenty of space if their plumage is to remain in a good condition. They can easily be associated with other birds, even with pheasants.

The two main species, the Indian (*cristatus*) and the Green (*muticus*), cross freely, producing completely fertile hybrids.

THE INDIAN PEAFOWL (*Pavo cristatus*) Plate VI

This is the universally known, common, bird that has been kept and reared outside its natural state so long that it has undergone a number of variations in size and build. It is important to note that

FIG. 47. White Peafowl.
(*Pavo cristatus alb.*)

because of the occurrence of the albino-factor an entirely new breed has come into being, the White Peafowl and the Pied Peafowl.

The White Peafowl is an albino of the original form and wholly white. Its beauty lies in the fact that, owing to the special structure of the barbs of the feathers, the pattern in the white colour has remained clearly visible. The Pied Peafowl shows white patches in its normal colour; they vary considerably, both in size and location.

A different, very interesting, mutation resulted in the Black-winged Peafowl (*Pavo cristatus nigripennis*). It is not known when this mutation first appeared, though Latham mentions this bird as early as 1823. The Black-winged Peafowl is truly magnificent (more beautiful than the original Indian birds) and just as easy to keep and to rear as the other Peafowls.

All the Peafowls described here are perfectly hardy.

The Green Peafowls are found farther east than the Indians, and live in south-eastern Assam, Burma, Siam, Indo-China, Malaya and Java. Curiously enough, they are found neither in Sumatra nor in Borneo.

The Green Peafowls are also even more beautiful in colour than

FIG. 48. Green Peafowl.
(*Pavo muticus muticus*)

their Indian relations. At the same time their legs are longer and the female more closely resembles the male, whereas in the Indian Peafowls there is a marked difference between the sexes. For one thing, the hen also has spurs on the tarsi, but her legs are much shorter. The hen and young males can be distinguished without too much difficulty through the loral patch between the eye and the bill: it is brown in the hen and bluish black in the cocks.

In their habits the Green Peafowls strongly resemble the Indian, though they are usually rather wilder and more wary. They do very well in the aviary or in a park, and they breed easily. They are less hardy than the Indian Peafowls, and should be sheltered against frost. A closed night pen is sufficient; heating is not essential. The aggressive nature of the cocks unfortunately makes it impossible to keep several at liberty in a park or in the same pen.

Sometimes their bad tempers may turn them against their keeper, and they can inflict much pain with their sharp, long spurs. The following are pure-breeding subspecies of the Green Peafowls:

The Javanese Green Peafowl (*Pavo muticus muticus*);
The Indo-Chinese Green Peafowl (*Pavo muticus imperator*), which is the generally known and most commonly kept form; and
The Burmese Green Peafowl (*Pavo muticus spicifer*). Plate VI

Several members of the subfamily "Phasianidae" have been left out altogether. A choice had to be made out of the available material, and I have been guided in the first place by the thought "which birds come into consideration because of their frequency and their price". All the same, this choice does not imply that all pheasants discussed are available in ample quantities and may be obtained at little cost—far from it. On the other hand, it can be said that all the birds that have not been included are extremely rare and highly expensive. But they are also of great beauty. As an example I might mention the Blood Pheasants (*Ithaginis*—thirteen species) or the Koklass (*Pucrasia*—ten species) and finally the newly discovered and incredibly rare Congo Peacock (*Afropavo congensis*).

Extensive descriptions of these birds would lead us too far afield and transcend the limits of this work, which only aims at propagating the keeping and rearing of the Pheasant.

THE ORNAMENTAL PHEASANT TRUST

The Ornamental Pheasant Trust was founded in 1959, and a collection started at Great Witchingham, Norfolk, quickly developed into the most comprehensive in Britain, and one of the largest in Europe.

The following are its objectives:

(*a*) To acquire and maintain a representative collection of Phasianidae under suitable conditions for scientific study and research.

(*b*) To protect and save from extinction all species of Phasianidae (and particularly the rarest) by propagation in captivity and especially to encourage their preservation in the wild by such means as are available including, where practicable, re-introduction of any species to its native land.

(*c*) To sponsor scientific study of these species in every part of the world both in captivity and in the wild.

(*d*) To endeavour to evolve improved methods and conditions for the keeping in captivity of all species of Phasianidae.

(*e*) To enable the public to see as many species of Phasianidae as possible, thereby fostering their interest in these birds and their problems of existence and demonstrating the importance and relevance of aviculture: (i) for their education as well as pleasure; (ii) for purposes of scientific research; and (iii) to save threatened species from extinction.

(*f*) To establish and support or aid in the establishment or support of any educational charitable purpose in any way connected with the foregoing purposes.

(*g*) To diffuse information on any branch of aviculture with particular reference to the objects aforesaid in such manner as is thought fit.

Mr. Philip Wayre, the Director of the Trust, states:

"If some species are to survive, positive steps must soon be taken to propagate their numbers in captivity and to carry out further research into their lives and habits. If this can be done successfully it may be even possible to re-introduce a species where it has become extinct or extremely rare in its natural range.

"To achieve these aims, ornithology and aviculture must go hand in hand, and it is regrettable that, in the past, the ornithologist has often overlooked the aviculturist when the latter would have been a powerful ally. The Whooping Crane of America, a species very close to extinction, immediately comes to mind. The Nene goose has fared better thanks to the wonderful work the Wildfowl Trust has done in building up the numbers of these birds at Slimbridge.

THE ORNAMENTAL PHEASANT TRUST

"Members of the pheasant family comprise some of the most beautiful and interesting birds in the world. It is to ensure their continued existence that the Ornamental Pheasant Trust has been formed."

Membership of the Ornamental Pheasant Trust is on a basis of:

(*a*) Ordinary Membership
(*b*) Sustaining Membership
(*c*) Life Membership

and the address is:

The Ornamental Pheasant Trust,
Hawks Hill,
Great Witchingham,
Norwich,
Norfolk.

THE JEAN DELACOUR AMERICAN GAME BIRD PARK AND PROPAGATION CENTRE

The Jean Delacour American Game Bird Park and Propagation Centre includes all species of game birds, with particular emphasis on the preservation in captivity of pheasants and waterfowl. It is an institution sponsored by the two national ornamental game-bird associations of America, the International Wild Waterfowl Association and the American Game Bird Breeders' Cooperative Federation. It is in Salt Lake City, Utah, an ideal area for the propagation of all types of pheasants, waterfowl, partridge, quail, etc.

The purpose of this game-bird park and propagation centre is to establish game birds in captivity, and especially the rarer varieties. Membership dues, donations, admittance fees are used for the development and maintenance of the Centre, and for purchasing and importing rare species, etc.

This project is dedicated to the preservation of game birds in captivity, and a comprehensive display of many species of the world's game birds.

The Centre is named after the world-famous ornithologist and aviculturist, Jean Delacour. The Director is George A. Allen, Jr. Membership is extended to everyone interested in helping to preserve game birds, and the mailing address of the Centre is:

1328 Allen Park Drive,
Salt Lake City,
Utah, U.S.A.

LITERATURE

Game-birds of India, Burma and Ceylon, by E. C. Stuart Baker
A Monograph of the Pheasants, by W. Beebe
Pheasants, their Lives and Homes, by W. Beebe
Monografia dei Fagiani, by Prof. A. Ghigi
Fagiani, Pernici and altri galliformi da caccia and da voliera, by Prof. A. Ghigi
Pheasants in Covert and Aviary, by T. Townsend Barton
Fasanen, by Dr. C. von Wissel and M. Stefani
Pheasants, by W. B. Tegetmeier
Game Bird Breeders, Pheasant Fanciers and Aviculturists' Gazette
Ibis
Avicultural Magazine
A Monograph of the Pheasants of Japan, by N. Kuroda
Notes on Some Pheasants, by Marquess M. Hachisuka
Game Birds of India, by E. W. Oates
The Pheasants of the World, by Jean Delacour